Journal Bearings in Turbomachinery

Journal Bearings in Turbomachinery

D. M. SMITH

D.Sc., LL.D., F.I.Mech.E., F.R.Ae.S., F.R.S.

Lately Consulting Mechanical Engineer,
Turbine-Generator Division,
Associated Electrical Industries Ltd.

CHAPMAN AND HALL LTD

11 NEW FETTER LANE, LONDON EC4

First published 1969

Printed in Great Britain
by Richard Clay (The Chaucer Press), Ltd.,
Bungay, Suffolk

SBN 412 09590 4

Distributed in the U.S.A.
by Barnes and Noble, Inc.

Contents

APPENDICES

Preface

This book deals with the functioning of hydrodynamic journal bearings in turbomachinery. It makes particular reference to large turbine-generator and marine propulsion plant.

Journal-bearing design in this field has been based mainly on experience supplemented by full-scale experimental test. Development is becoming influenced to an increasing extent by research and analysis. Particular attention is given in this book to correlation of research and analytical work with the observed operating characteristics of journal bearings. The physical phenomena in bearings are complicated, and analysis is rendered convenient only by making simplifying assumptions. The engineer must know which assumptions are serviceable and in what operating conditions they may be applied.

Current British and European practice in journal bearings is illustrated. An examination is made of steady running characteristics, as predicted by theory and as established by test. Some account is given of the dynamic characteristics of journal bearings and of their influence in machine vibration. Service experience of journal bearings is reviewed, and reference is made to possible future trends in development.

The book is the outcome of work on turbine plant with Metropolitan-Vickers and its successor Associated Electrical Industries. The A.E.I. and English Electric activities in this field have recently been incorporated in English Electric–A.E.I. Turbine-Generators Ltd. The author expresses his gratitude to the Company for permission to publish the results. He thanks the English Electric Co. Ltd., C. A. Parsons & Co. Ltd., A.E.G., and Alsthom for information on their bearing practice so generously supplied.

He gratefully acknowledges permission from the Institution of Mechanical Engineers to reproduce material from papers to the Institution.

Many of the author's colleagues at A.E.I. have made contributions to the investigations recorded. It is impracticable to acknowledge all by name, but he would express his special thanks to Dr A. Frankel and Dr A. W. Crook for encouragement and facilities in the preparation of the book.

D. M. Smith

Notation

The analytical relations of the book apply with any consistent system of units.

Dots over symbols indicate differentation with respect to time.

The entry under 'Chapter' indicates the chapter in which the symbol is defined. The leading dimensions for a circular journal bearing are shown in figure 4.1. The notation for investigation of dynamic characteristics is illustrated in figures 9.1 to 9.4.

A few symbols used only locally are not included below.

Symbol		*Chapter*
B_u, B_v	Apparent force – velocity coefficients	11
C	Diametral clearance in circular bearing; twice difference in radius between journal and bottom land in non-circular bearing	4
D	Diameter of journal	4
E	Nondimensional energy input in moving the axis of vibrating journal once round an elliptical orbit	9
H	Power loss	4
H_p	Petroff loss	4
J	Mechanical equivalent of heat	5
K_j	Power loss coefficient	5
K_q	Oil flow coefficient	5
K_u, K_v	Apparent force-displacement coefficients	11
L	Bearing length	4
N	Speed of rotation, revolutions per unit time	4
N_T	Taylor number	5
P	Nondimensional force component	5
P^*	Actual force component	5
P_x, etc.	Suffix indicates direction of force component	5
Q	Oil flow through bearing	4

Symbol		*Chapter*
Q_R	Reference oil flow	4
R	Radius of journal	4
(Re)	Reynolds number	5
S	Duty parameter	4
U	Peripheral velocity of journal	4
U'	Peripheral velocity taking transverse motion of journal into account	9
V	Reference velocity	5
W	Steady load transmitted through bearing	4
X, Y	Amplitude of journal vibration	9
a_{11}, a_{22}, etc.	Force–displacement coefficients	9
b_{11}, b_{22}, etc.	Force–velocity coefficients	9
c_{11}, c_{22}, etc.	Force–acceleration coefficients	10
c_p	Specific heat of oil	5
e	Eccentricity of journal	4
f	(Chapters 5, 6, 7) Friction factor	5
f	(Chapter 9) Frequency	9
g	Gravitational acceleration	11
h	Nondimensional film thickness	4
h^*	Actual film thickness	4
j	Calculated power loss coefficient	4
k, l	Nondimensional coordinates along and normal to attitude line	4
l_c	Reference length	5
m, m'	Exponents	5
m	$(1 - n)$	App. IV
n	Eccentricity ratio	4
p	Nondimensional pressure in film	4
p^*	Actual pressure in film	4
p'	Mean pressure on projected area of journal	4
p_k, p_l, p_m	Nondimensional mean pressures	4
q	Calculated oil flow coefficient	4
r	Radial clearance	4
s	Nondimensional coordinate in tangential direction	4
s^*	Actual coordinate in tangential direction	4
t	Temperature rise of oil	5
t_1	Oil-inlet temperature	5
t_2	Mean oil-outlet temperature	5
t	(Chapter 9 and App. VII) Time	9

Symbol		*Chapter*
u, v	Nondimensional coordinates in directions of principal axes of elliptical orbit	11
x, y	Nondimensional coordinates in horizontal and vertical directions	4
x^*, y^*	Actual coordinates in horizontal and vertical directions	4
z	Nondimensional axial coordinate	4
z^*	Actual axial coordinate	4
Ω	Nondimensional speed of rotation	11
β	Angle from attitude line	4
β_1, β_2	Angles at beginning and end of pressure-bearing film	4
δ_1, δ_2	Phase angles	9
ϕ	Attitude angle	9
γ	Tangent angle of attitude locus	9
μ	Viscosity of oil	4
ν	Kinematic viscosity of oil	App. V
ρ	Density of oil	5
ω	Angular speed of rotation	4

PART ONE

Theory and Practice

CHAPTER ONE

Bearing Development and Bearing Theory

Journal bearings had centuries of development by empirical methods before there was any realization of the nature of the physical action by which a hydrodynamic lubricating film is formed. Empirical methods still retain an important place in journal-bearing development.

The build up of pressure in the film was observed by Beauchamp Tower in 1883. Three years later Osborne Reynolds explained the build up of pressure by viscous action in a convergent film [1]. The physical understanding thus obtained assisted designers in choosing the position of oil-supply and drain grooves in journal bearings. The first analytical investigation of loss in a hydrodynamic journal bearing was published in Russia, by Petroff, in 1883.

Much analytical work on lubrication in bearings was subsequently published, but for many years engineers found this analysis of little relevance to journal-bearing design. Application of Reynolds' theory led to revolutionary advance in thrust-bearing design with the invention of the tilting-pad thrust bearing by Kingsbury and by Michell independently. The theory has not led to any similar spectacular advance in journal-bearing design.

The progressive development of turbomachinery since the 1880s has led to repeated requirements for journal bearings to operate at increased peripheral speed and increased load. The peripheral speed requirements exceed those in large bearings for other applications. The advances made in bearing design were based on service experience, supplemented from time to time by full-scale rig tests of new designs. These advances were facilitated by an increasing range of research which contributed to improvement in background knowledge.

A general review of bearing practice, experimental work, and lubrication theory in 1937 was presented at the I.Mech.E. General Discussion on Lubrication [2]. The designers of turbine journal bearings were at that time groping with analytical theories which gave results inconsistent with one another and with experience.

It is useful to consider why empirical development of journal bearings has persisted for so long, while analytical theory was late in being of real service to engineers.

The Reynolds relation for pressure variation in the film has a simple physical basis, and it applies with accuracy so long as there is stable laminar flow in the pressure-bearing film. The calculation of journal-bearing characteristics from this relation, however, presented great difficulties. The extent of the pressure-bearing film was not known. Analytical theory is manageable only when it is based on simple assumptions which imperfectly represent the physical realities in turbine journal bearings. The analytical and experimental work of many investigators has led in course of time to general acceptance for journal-bearing calculation of a conventional hydrodynamic theory of laminar flow. The limitations of the conventional assumptions, as discussed later, must be borne in mind when applying this conventional theory.

By 1950 the theory was giving results in reasonable accord with the observed steady running characteristics of many turbine bearings. There were, however, some anomalous results, particularly on large high-speed bearings. Most of these anomalies have been explained by the occurrence of nonlaminar flow in the bearings. The serious effects of nonlaminar flow, especially in increasing bearing loss, have been accentuated in recent years by the use of bearings of still higher peripheral speed.

A contributory cause of the persistence of empirical development is that in hydrodynamic journal bearings the design is not determined primarily by the normal operating conditions at full speed, full load. The conditions in starting and at low speed may be the limiting ones, and liability to failure at normal speed depends on the occurrence of abnormal conditions, such as dynamic loading in emergencies, misalignment, or deficiency in supply or cleanliness of oil. The designer requires to know the bearing performance at full speed, but this is for the purpose of determining loss and designing the lubricating-oil system.

In the running of turbomachinery it has long been known that the journal bearings greatly influence machine vibration, both in response to disturbance (such as out-of-balance) and in stability of running. The problem was attacked analytically and experimentally, but engineers were perplexed by discordant theories of the dynamic phenomena in the bearings, and by difficulty in distinguishing the dynamic effects of the bearings from the dynamic effects of the bearing supports. Acceptance of a conventional theory of the dynamic characteristics of journal bearings became widespread about 1960. This theory, which is based on the conventional theory of characteristics in steady running, was also the outcome of the cumulative work of many investigators. The conventional theory of dynamic characteristics applies to journal motion of small amplitude in a bearing with laminar flow in the film. Understanding of the dynamic characteristics of journal bearings is still far from complete.

The build-up of knowledge of lubrication from analytical and experimental contributions has been summarized by Mayo D. Hersey [3]. Contributions to knowledge continue to be made at an increasing rate. A review of recent research and its application to design was undertaken by the I.Mech.E. Conference of 1967 on Lubrication and Wear [4].

CHAPTER TWO

Current Practice in Journal Bearings

Standardization of journal bearings

It is advantageous from the production point of view for a manufacturer to use one standard design of journal bearing, made in a small number of standard sizes. The degree of standardization attained is, however, limited by other considerations. The economic benefits of reduction of machine dimensions and of reduction of bearing loss justify the use of bearings of small steps in diameter and of more than one length in each diameter. The operating conditions of turbine plant also affect bearing design in various ways. A bearing design which is well suited to a direct-coupled turbine generator is not suited to a journal which reverses in direction, or to a bearing where the line of application of the load swings through a large angle. Journal-bearing design is frequently influenced by the effect of the bearings on various types of rotor vibration. Manufacturers who produce a wide range of turbomachinery therefore use bearings of more than one design.

Journal bearings for turbine-generator plant

In direct-coupled turbine plant for electricity generation the direction of rotation is always the same, the normal operating speed is constant, and the steady load carried by each bearing is determined mainly by the weight of the rotor. In a multi-bearing set there may be redistribution of the 'steady' load between bearings due to change in bearing alignment. Rapidly fluctuating loads, associated with machine vibration, are superimposed on the steady load. In good running these fluctuating loads are smaller than the steady load.

Figure 2.1 shows a bearing design used by A.E.I. in steam and gas turbines. The bearing surface is of white metal, carried in a steel liner. The bearing shell is in halves, which are shown separated in the perspective drawing. In normal operation the journal (which is not shown) and the bearing are separated by an oil film. The steady load is applied vertically downward from journal to bearing. The bearing shell is carried in a housing of the type shown in a later figure.

Lubricating oil is supplied through a channel and axial groove at the horizontal joint on the rising side of the bearing. The oil flows round a shallow groove in the top half of the bearing. This groove is bounded at each side by strips of bearing surface which limit axial leakage and which provide bearing areas to restrain the journal in the event of movement upwards. The oil stream passes through an axial groove near the horizontal joint on the falling side of the bearing. The narrow oblique grooves from the bottom corners of this axial groove are provided to drain out solid particles in the oil which might otherwise pass into the film.

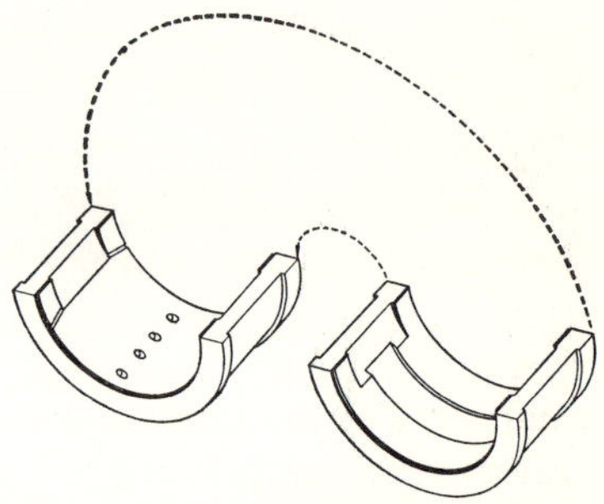

FIG. 2.1. Turbine bearing – A.E.I.

The oil enters the lubricating-oil film on the bottom half of the bearing. The lead-in from groove to film is carefully rounded. This film sustains the main bearing load. Some of the oil which has passed through the film escapes at the sides of the bearing. The remainder of the oil passes into the supply groove, where it mixes with the incoming oil and is recirculated through the bearing.

Holes are provided in the bottom of the bearing to admit jacking oil. The jacking oil is supplied under substantial pressure (about 1,000 lb/sq in.) at starting and in low-speed running. This high-pressure oil ensures formation of an oil film at very low speed when otherwise there would be metallic contact or imperfect film formation. In the absence of jacking oil an oil film separating journal and bearing is formed by hydrodynamic action above a low speed.

Bearings of this design are used with circular bore in the smaller sizes and with 'elliptical' bore in the larger sizes. The 'elliptical' bore is produced by machining a circular bore with shims in the horizontal joint of the shell and then assembling the bearing without the shims. A typical clearance ratio (ratio of diametral clearance to diameter) is 0·025 horizontally and 0·0015 vertically.

White metal is cast into the mild-steel liner to form the bearing shell. The white-metal thickness varies from 0·030 in. in the smaller to

0·050 in. in the larger diameters. The white metal commonly used is a babbitt of 85–89% tin content (No. 2 alloy of British Standard Specification 3332/1961).

The minimum acceptable journal diameter is fixed by stress or critical speed considerations external to the bearing. The bearing length is chosen for load on projected area not to exceed about 250 lb/sq in. Standard bearing diameters and lengths are used where possible. The standard lengths are 50 and 75 per cent of the journal diameter. This bearing design has been used with normal peripheral speed up to 250 ft/sec. The lubricant is medium-grade turbine oil supplied at about 10 lb/sq in., 120° F.

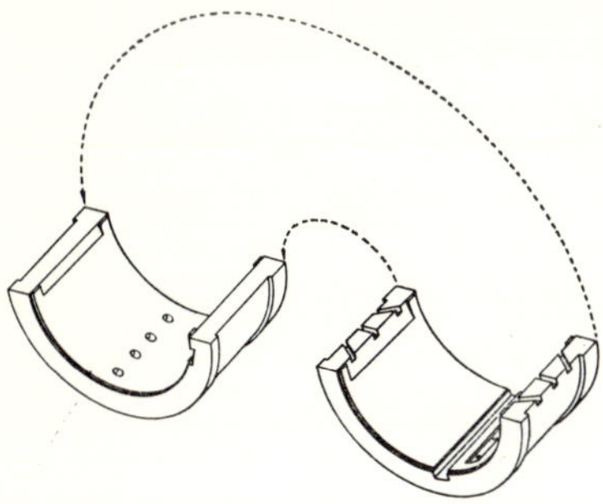

FIG. 2.2. Three-land bearing.

An alternative A.E.I. design – the three-land bearing – is shown in figure 2.2. This design is more resistant than the previous one to low-frequency whirl. It is used on generator bearings and on some turbine bearings. The bearing has three separate bearing surfaces, or lands, of equal width but subtending different arcs, one land in the bottom half and two in the top half. There is an oil-supply groove with lead-in at the beginning of each land, and a drain groove at the end. Recirculation of oil is limited by an axial strip of bearing surface between each drain groove and the following supply groove.

The three-land bearing is generally of circular bore with clearance ratio about 0·0013. Elliptical clearance is used in special cases. The materials of construction are as for the bearing previously described. The three-land design has been used with service loads and speeds similar to that design, and has been rig tested up to much higher loads and speeds.

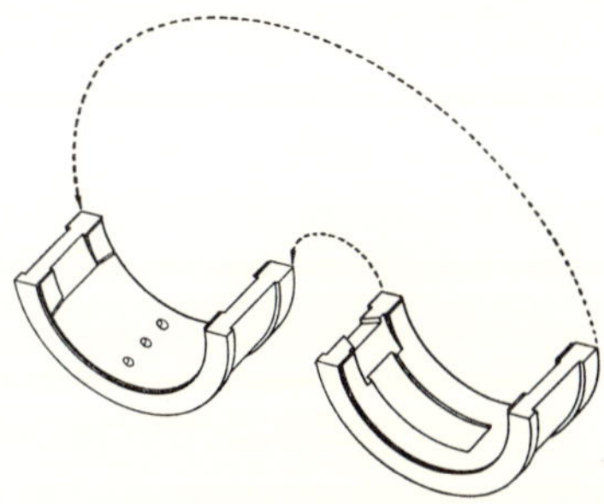

FIG. 2.3. Stabilized bearing.

A bearing design for protection against a different type of whirl is shown in figure 2.3. This is the stabilized bearing which is used to safe-

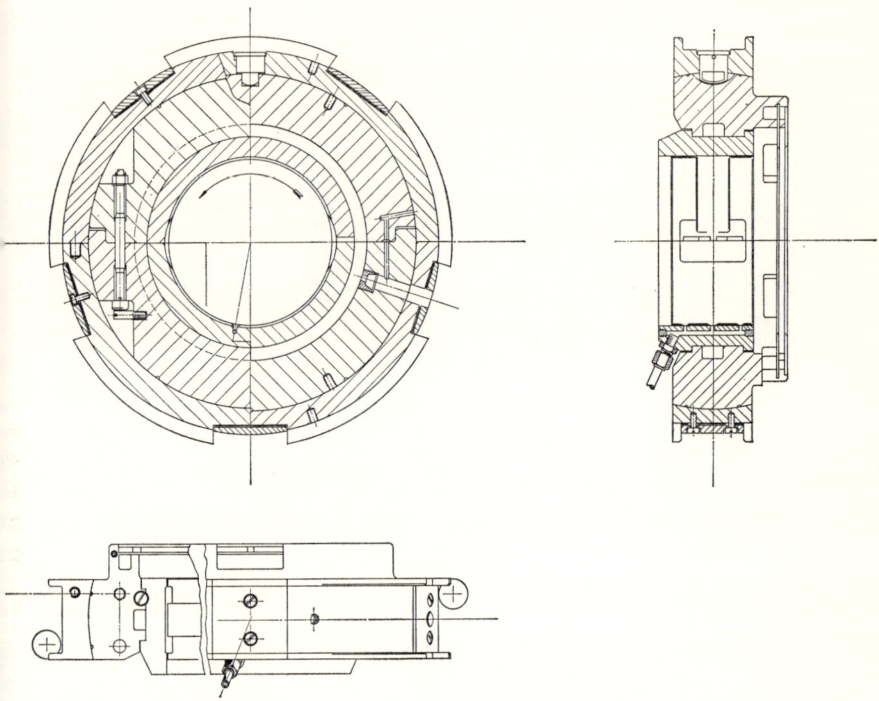

FIG. 2.4. Bearing assembly.

guard against light-load instability when the projected load on the bearing area is less than 60 lb/sq in. The design has also been found beneficial in some other cases.

Oil is admitted to the supply groove on the rising side of the bearing, and passes into the stabilizing groove in the top half. This groove is sharply cut off at 135° from the horizontal joint. Oil pressure is built up in the groove by rotation of the journal; this pressure increases the load carried by the oil film in the bottom half of the bearing. Oil escaping from the stabilizing groove lubricates the remainder of the bearing. There is an axial distribution groove on the falling side of the bearing; this groove has no external oil supply or drain connection.

This bearing design is used mainly on small sets, since low bearing loading occurs most frequently on these sets. The width of a stabilized bearing is usually half the diameter. Narrower bearings are used in certain cases. The bearing is of circular bore with clearance ratio depending on diameter and speed. Jacking oil is sometimes provided, but is often unnecessary with lightly loaded bearings of this type.

A typical A.E.I. turbine bearing assembly is shown in figure 2.4. The

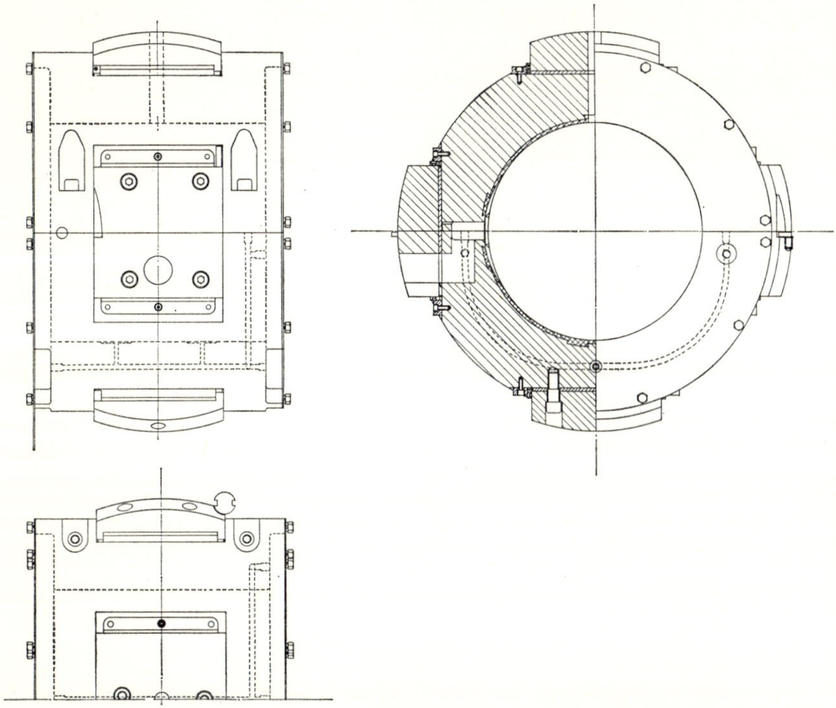

FIG. 2.5. Turbine bearing – Parsons.

liner is carried in a bearing which is in halves and is mounted in a spherical seat. The main oil supply enters through a recess encircling the liner, so that a liner with any design of oil inlet can be used in the same bearing. The high-pressure jacking oil is admitted through a separate supply pipe direct to the liner.

Spherical seating of journal bearings is normal practice except for small bearings of low length-to-diameter ratio. The primary function of the spherical seat is to facilitate bearing alignment during erection of plant. In service, owing to frictional restraint, the bearing is unlikely to move in its spherical seat under shaft vibration or stator distortion of small amplitude. Movement may take place in the spherical seat, particularly with a long bearing, if there should be large angular displacement of the journal relatively to the bearing pedestal.

A journal bearing used by C. A. Parsons and Company Ltd. is shown in figure 2.5. A significant difference in construction from the bearings

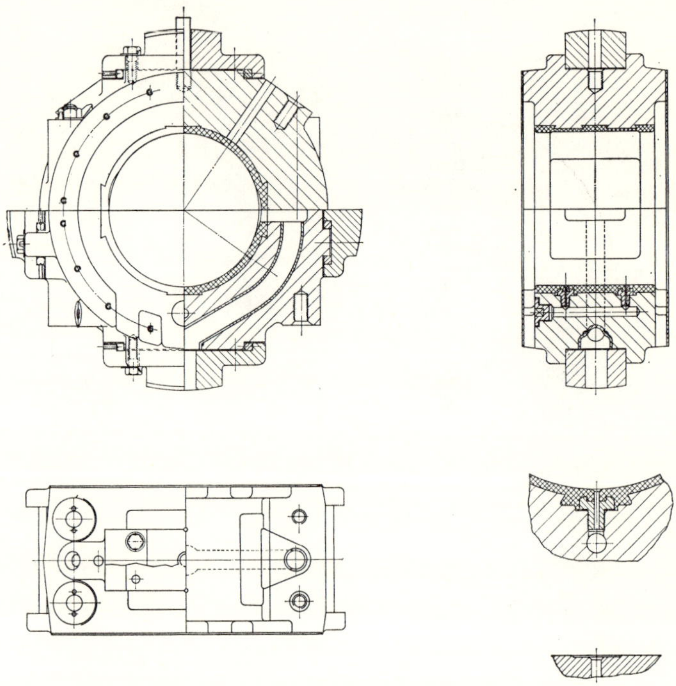

FIG. 2.6. Turbine bearing – English Electric.

previously described is that the white metal is cast directly into the bearing body and not into a separate liner.

The bearing is in halves, split at the horizontal centre line. It is held together by four bolts; location is provided by four dowels. The bearing is spherically seated in the pedestal on four pads, under which are shims to facilitate alignment. Secondary shims are provided under the side guides to maintain sphericity.

Oil is supplied at each side on the horizontal centre line. This oil enters through an axial recess subtending an angle of 90° at each side of the bearing. The oil drains at the ends of the bearing. High-pressure oil for starting is supplied to the bottom of the bearing. The bearing is of elliptical bore.

Typical applications involve bearing loads up to 215 lb/sq in. and peripheral speeds up to 250 ft/sq in. A bearing of this type has been rig tested to much higher loads and speeds. The normal lubricant is medium-grade turbine oil, supplied at 10–15 lb/sq in., 100–110° F.

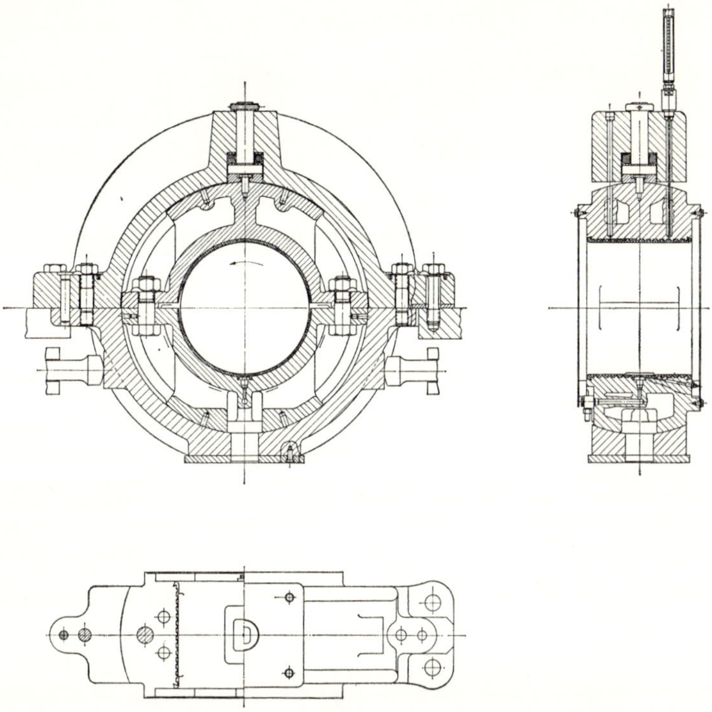

FIG. 2.7. Turbine bearing – A.E.G.

A large journal bearing of turbine type used by the English Electric Company Ltd. is shown in figure 2.6. Some comments on this design are made in ref. [5]. In this bearing the white metal is also cast directly into the bearing body. The bearing is in two halves, split at the horizontal joint, and is carried in a spherical seating. Oil is admitted at both sides. There are grooves in the bearing surface at each side and in the top half. The bore is cylindrical. Jacking oil is provided.

An example of German practice is provided by the A.E.G. turbine-generator bearing shown in figure 2.7. In this bearing oil is admitted at both sides through recesses on the horizontal joint to axial grooves. Jacking oil is admitted to a shallow groove in the bottom of the bearing. The white metal is cast into dovetail slots in the bearing shell. The bearing shell is usually in cast iron; some recent shells for larger bearings have been in cast steel. The shell is in halves, joined by two fitted bolts, and is spherically seated. To eliminate risk of undue distortion of the shell the reaction on the top spherical locating plate is applied through annular plate springs. Circumferential location of the shell is provided through the same locating plate.

Bearings of this type are used with load on projected area within the range 5–20 kg/cm^2 (71–284 lb/sq in.), with diameter exceeding 80 mm (3·15 in.) and with peripheral speed up to 75 m/sec (246 ft/sec). The length-to-diameter ratio is 0·6–0·9. The bearing is elliptical. The normal clearance ratio at 3,000–3,600 r.p.m. is 0·0023 horizontally and 0·0015 vertically. Smaller clearance ratios are used at lower speeds.

White-metal temperature is monitored by a thermocouple inserted in the bottom shell at an angle of 25° to the vertical in the direction of rotation. The temperature of the bearing shell is measured by a thermometer in an oil-filled pocket.

Figure 2.8 shows an Alsthom turbine bearing as an example of French practice. This bearing has wide axial grooves at both sides, interconnected by a shallow circumferential groove, of half the bearing width, round the top half. Oil is admitted on the rising side only through the recess at the horizontal joint. Jacking oil is admitted to a groove in the bottom of the bearing. Oil escaping at each end of the bearing is restrained by an oil baffle to flow through an outlet at the bottom. The white metal is cast directly into the bearing shell, which is in halves and is spherically seated. The bore is cylindrical. White-metal temperature is monitored by a thermocouple in a position similar to that in the A.E.G. bearing.

In all the bearings which have been described a main bearing surface on which an oil film is formed is provided in the bottom half. Additional bearing surfaces are provided in the top half. The oil-supply arrangements ensure that all supply and distribution grooves run full. Oil is available to form a film at each bearing surface. There is thus a general resemblance between these bearings, although there are substantial variations in detail.

Tilting-pad journal bearings have also been used in turbine-generators, but to a greater extent in American than in British or

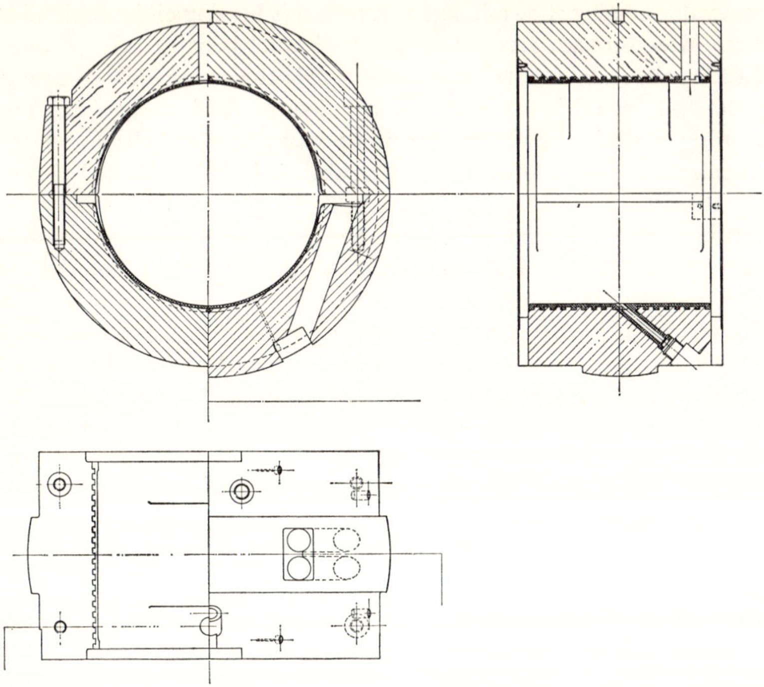

FIG. 2.8. Turbine bearing – Alsthom.

European practice. They generally have four or more pads distributed round the circumference. Each pad is carried by a pivot or flexible support. The bearing is arranged for oil supply to the inlet edge of each pad.

The lubricating-oil system on turbine-generator plant is normally common to all the journal and thrust bearings and the relay gear. Oil is circulated through an oil cooler by a pump driven by the main turbine. Separate motor-driven pumps are used for jacking oil and for stand-by supply.

Journal bearings for marine gearing

Journal bearings for marine gearing differ functionally in two important respects from the journal bearings previously described – the direction of rotation reverses, and the applied load varies widely in magnitude and direction. Prolonged running may take place at any speed from very low speed to full speed. The load is usually low at starting and increases in magnitude with rising speed. During reversal there is a continuation of high load at low speed owing to the propeller torque from the ship's speed.

Much experimental work has been carried out in the course of development of bearings for marine gearing of high performance. Two examples are given of A.E.I. bearing designs which have been adopted for these operating conditions.

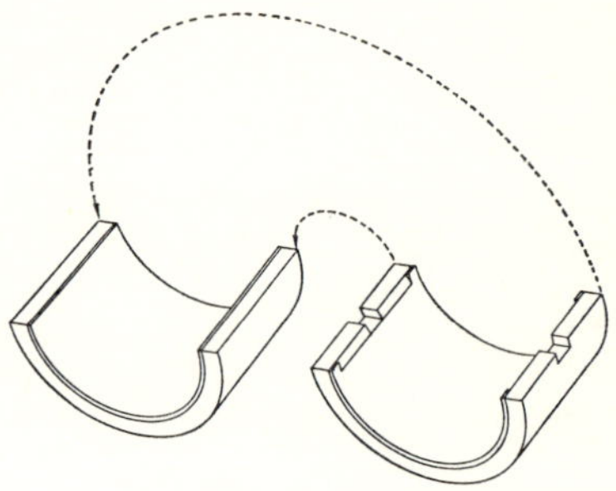

FIG. 2.9. Pinion bearing.

Figure 2.9 shows a bearing shell for the pinion line (the high-speed line) of a marine reduction gear. The shell is in halves, with oil admitted on each side at the horizontal joint. The oil is distributed through axial grooves extending over most of the bearing width. The edges of the grooves are blended into the bearing surface; the surfaces in the top and bottom halves are otherwise unbroken. The bearing is of circular bore with clearance ratio 0·002. This bearing has been used in Naval service with load on projected area up to 500 lb/sq in. at peripheral speed 225 ft/sec. Under such conditions the proportions of the bearing and the clearance ratio are important.

Figure 2.10 shows a bearing shell for the main gearwheel in the low-speed line. One of the problems presented in this case was the requirement of carrying substantial load at low peripheral speed, since propeller r.p.m. is low by turbine standards.

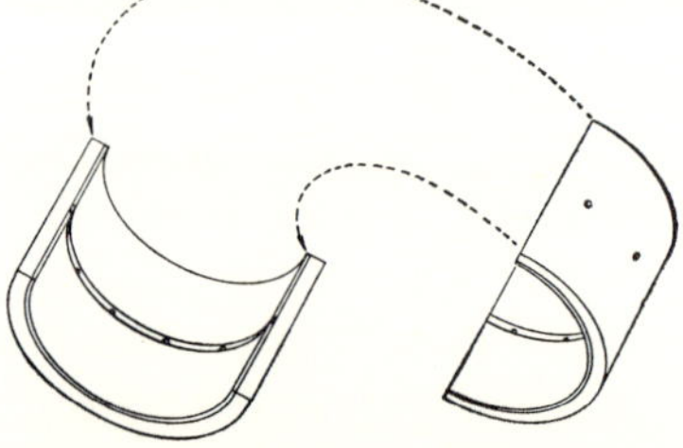

FIG. 2.10. Main gearwheel bearing.

The shell is in halves, with the joint set at 45° to the horizontal, so that the joint is removed from the direction of highest load. Oil is admitted through a continuous circumferential groove in the middle of the bearing. The edges of the groove are rounded at inlet to the oil film. The bearing surface is of cylindrical bore and is continuous at each side from centre groove to end of the bearing. The oil from the film drains through the ends.

The bearing was developed on rig test to carry maximum load 345 lb/sq in. at maximum peripheral speed 15 ft/sec.

In both the marine gear bearings described the shells are of steel lined with tin-base white metal. The shells are mounted in the gearbox.

PART TWO

Performance in Steady Running

CHAPTER THREE

Functioning of Journal Bearings

When a journal bearing operates under conditions of hydrodynamic lubrication there is no contact between the rotating journal and the fixed bearing surface. In steady running the journal assumes an eccentric position, described as the journal attitude, in which the journal centre is displaced from the centre of curvature of the bearing surface in such a way as to form a convergent film of lubricant underneath the load. Pressure is built up in the convergent film by the journal rotation.

At starting and at very low speed a lubricating film is not formed by hydrodynamic action. When there is no external supply of high-pressure lubricant to lift the journal conditions of boundary lubrication prevail, with surface contact or very close approach to surface contact between journal and bearing. Prevention of surface damage in these conditions involves proper choice of materials and lubricant, as well as of design and of finish of surfaces and of bearing loading.

The phrase 'boundary lubrication' has been in use for a long time without clear understanding of what physical actions it covers. Recent investigations of these actions are discussed by D. Tabor [6]. Elasto-hydrodynamic action plays a considerable part, with local elastic distortion of the bearing surface comparable in magnitude with the thickness of the lubricating film formed. Experience has led engineers to suitable choice of materials and lubricants. The occurrence of boundary lubrication in the bearings of turbomachinery has become less frequent with the general use of jacking oil at very low speeds, but it remains necessary to choose materials and lubricants which have good properties in boundary lubrication.

Lubricating oil, as normally used in turbine plant, is a fluid of moderate cost which has good properties both in boundary lubrication and in hydrodynamic lubrication. In this and later chapters the word 'oil' will frequently be used to indicate the lubricant even where the context would admit other liquid lubricants.

Gas can be used as a lubricant, but its functioning in journal bearings differs from that of oil. The differences include substantial compressibility effects and absence of broken film, in addition to great change in the physical properties of viscosity and density which control behaviour in the lubricating film. Gaseous lubrication is not discussed in this book except where specific reference is made to it.

In the absence of jacking oil the hydrodynamic film in a journal bearing forms at a low speed. With jacking oil, hydrodynamic action plays an increasing part as the speed rises. The following description refers to a bearing in which the oil film is formed purely by hydrodynamic action. Sectional diagrams of lubricated bearings are shown in figures 4.1 and 4.3 of the next chapter.

Lubricating oil is supplied to the bearing through one or more supply channels or grooves. Some of the numerous grooving arrangements used have been described in the previous chapter. Oil escapes from the ends of the bearing and also from drain grooves when provided. The oil-drain space is normally at atmospheric pressure.

The continuous oil film in which pressure is built up is known as the pressure-bearing film. The oil film is unable to sustain a pressure much below atmospheric. Where the pressure is sufficiently low, the film breaks or cavitates. The pressure in the broken film is usually in the neighbourhood of atmospheric pressure.

The pressure-bearing film starts either at an oil-supply groove or near the position where the clearance becomes convergent, provided that sufficient oil is available to form the film. Pressure-bearing films may be formed separately in various parts of a grooved bearing or of a bearing which is not of circular bore. The journal load is equal to the resultant of all the radial forces on the journal, including any forces exerted by pressure in oil grooves.

When tilting-pad journal bearings are used a pressure-bearing film is formed on each pad. The pad tilts so that the reaction from the film passes through the pivot.

In many turbine journal bearings steady loading is provided by the rotor weight only; in such cases the bearing load is constant and vertical. Turbines with partial admission have bearing loads which are affected in magnitude and direction by the steam-admission conditions. Gear bearings have loads which are greatly affected by the reactions at gear teeth. In running, as already mentioned, unsteady loads due to various causes are superimposed on the steady load.

The oil flow through a bearing is determined by the bearing design and operating conditions. When the oil supply is at atmospheric pressure the pumping action of a well-designed bearing is sufficient to pass an adequate oil quantity. It is normal practice to supply oil to the bearing at a pressure above atmospheric. In some cases the oil is supplied through a restricting orifice in the oil inlet; this may be used to regulate oil distribution between bearings in a common lubricating-oil supply system.

The oil is raised in temperature by the work done as it passes through the bearing. In turbine practice the inlet-oil temperature and the outlet-oil temperature are usually observed by instruments permanently installed. The outlet-oil temperature results from the mixing of oil streams at different temperatures, and does not bear any simple relation either to the maximum oil temperature in the film or to the maximum temperature of the bearing metal. Nevertheless, in the light of experience for a bearing of familiar design the mean outlet temperature can be used as a criterion of safe operation. The mean outlet temperature is also used in bearing-performance calculation.

CHAPTER FOUR

Conventional Hydrodynamic Theory of Laminar Flow

Assumptions for conventional calculation

In real bearings the hydrodynamic conditions are complicated and are not precisely known. The variation in viscosity of oil within the bearing is large, imperfections of workmanship and temporary distortion of the bearing surface may be appreciable compared with the film thickness, the conditions determining the beginning and end of the pressure-bearing film are not accurately defined, and the film flow regime experienced in bearings of moderate speed does not prevail in all bearings. Hydrodynamic calculation is therefore made for a reference bearing in which the assumed conditions are simplified as compared with those in the real bearing.

The conventional assumptions for calculation of performance of the reference bearing include:

(1) Film thickness everywhere small compared with journal dimensions.
(2) Journal cylindrical and of specified diameter; bearing surface to simple geometrical specification without local distortion.
(3) Journal axis parallel to bearing axis.
(4) Inertia of oil in film negligible.
(5) Oil film unable to sustain sub-atmospheric pressure.
(6) Oil pressure atmospheric along the supply and drain boundaries of the pressure-bearing film, and also throughout the film-clearance region outside the pressure-bearing film where the oil film is broken or cavitated.
(7) Laminar flow in the pressure-bearing film.
(8) Viscous shearing loss in the clearance region outside the pressure-bearing film, this space being taken as partly filled with oil.

(9) No contribution to loss from oil in grooves or drain spaces adjoining journal.

(10) Oil a simple Newtonian liquid with viscosity independent of shear rate.

(11) Viscosity and density of oil constant throughout the bearing.

It is also assumed for the present that the journal is running steadily under constant load.

Validity of the conventional assumptions

A number of investigations, analytical and experimental, have been made of the validity of the conventional assumptions for journal-bearing calculation and of the effect of the departures from these assumptions which occur in real bearings. Brief notes are given below on the conclusions reached so far as they apply to turbine-type bearings in steady running.

(1) The assumption of small film thickness is always valid.

(2) The workmanship and finish of both journal and bearing must be of very high standard if the accepted hydrodynamic theory is to be reasonably accurate, and local distortion of the bearing due to strain or thermal expansion must be very small. The effect of departure from the assumed rate of convergence of the film thickness is of special importance.

(3) The journal axis is rarely parallel to the bearing axis. Within certain limits, the lack of parallelism normally experienced has only a moderate effect on the loss and on the mean journal attitude, but it may diminish the minimum film thickness to a serious extent.

(4) Analytical investigation has shown that inertia of the oil has generally little effect on flow in the film at speeds which give stable laminar flow. Inertia of the oil must be taken into account in investigating stability of the laminar flow regime and characteristics of other flow regimes. Inertia has sometimes appreciable effect on bearing characteristics, even with laminar flow, when the oil is rapidly accelerated at entry to the film.

(5) The assumption that the oil film is unable to sustain sub-atmospheric pressure is not accurate. Oil films are able to sustain pressure below atmospheric; the pressure at break-up

is affected by the properties and condition of the oil, and cannot be specified in any simple way. In the larger bearings in turbomachinery the mean load on the projected area of the bearing is 100 lb/sq in. or more, and with this load, break of the film at, say, 2 lb/sq in. below atmospheric pressure makes little difference to the steady-running characteristics as compared with break at atmospheric pressure. With a lightly loaded bearing, such as is sometimes used on small-scale rig test, the break-up condition may substantially affect the characteristics as compared with those given by break at atmospheric pressure.

(6) The assumption that oil pressure is atmospheric along the supply and drain boundaries of the pressure-bearing film is usually incorrect at the oil-supply groove. Variation in oil-supply pressure has substantial effects on bearing performance; these effects are examined later. Assumption of supply at atmospheric pressure is the most convenient reference condition. Experimental evidence shows that the clearance region outside the pressure-bearing film normally contains a broken or aerated oil film at approximately atmospheric pressure; but sometimes there is local departure from atmospheric pressure in this region.

(7) Laminar flow holds in the pressure-bearing film from the speed at which a hydrodynamic film is formed up to a much higher speed at which laminar flow becomes unstable. In the smaller turbine bearings the flow remains laminar over all normal operating conditions. In the larger bearings (including nearly all bearings on recent turbomachinery of high rating) the flow is nonlaminar at normal operating speed. The conventional theory is inapplicable with nonlaminar flow.

(8) The assumption as to loss in the clearance under bearing surfaces outside the pressure-bearing film is approximately consistent with experience when the flow in the pressure-bearing film is laminar.

(9) Contribution to loss is made by oil in grooves, and also by oil in drain spaces unless the oil drains away immediately. The flow conditions in grooves are complicated. The groove-loss contribution is appreciable in some journal bearings, but is generally much smaller than the film loss.

(10) In normal journal-bearing practice, excluding starting and

very low-speed conditions, lubricating oil behaves as a Newtonian liquid with viscosity constant at given pressure and temperature. In starting and boundary lubrication conditions, with very thin films and local regions of high pressure, there may be departure from Newtonian characteristics.

(11) The viscosity of the oil varies widely between the oil-inlet temperature and the highest oil temperature in the bearing. Viscosity falls rapidly with rising temperature. The oil is heated by bearing loss. The variation of temperature in the film is limited by oil recirculation in the bearing, and is modified by exchange of heat with rotating journal. The viscosity variation is relatively moderate in the portion of the film which has great influence on the overall bearing characteristics. The overall characteristics of bearings in turbomachinery are adequately represented in many cases by assuming an 'effective' viscosity for the calculations of conventional theory. For some bearings this 'effective' viscosity may be taken as represented by viscosity at the mean oil-outlet temperature.

The viscosity of oil rises under high pressure. Viscosity increase due to increased pressure makes little difference to bearing operation in normal running; but it expedites formation of a hydrodynamic film in starting under load, when pressures are high at local asperities or in a limited arc of very small clearance. The density of the oil is also a function of temperature and pressure; but density varies little and hardly affects the bearing characteristics so long as there is no cavitation.

It thus appears that in bearings for turbomachinery a number of the conventional assumptions are approximately fulfilled, and in many cases the departure from fulfilment of others does not seriously invalidate the overall results of the conventional theory. Departure is liable to be serious when the film-flow regime is nonlaminar.

In assessing bearing characteristics by application of the conventional hydrodynamic theory, the engineer must keep in mind the imperfect correspondence of the reference assumptions with the real operating conditions.

Journal-bearing analysis

The analysis by which these conventional assumptions are applied in calculation of the steady running characteristics of journal bearings is given here in outline. For fuller treatment, reference should be made to textbooks [3, 7, 8, 9].

The bearing diagram, figure 4.1, and the diagram of oil-film section,

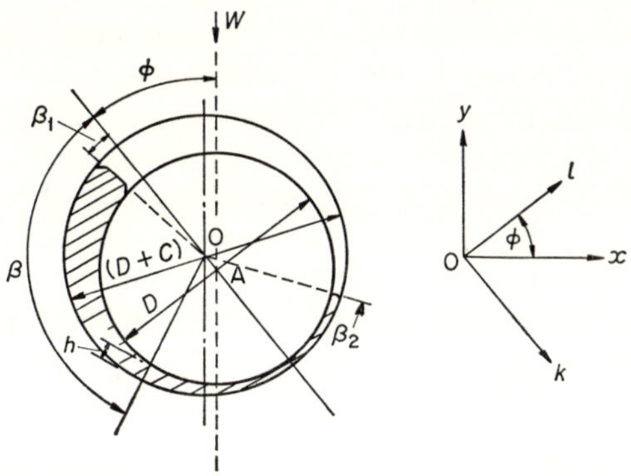

D = Journal diameter.
C = Diametral clearance.
L = Bearing length.
W = Steady load (applied vertically).
$P\omega = \frac{W}{DL}$ load on projected area of journal.
N = Speed of rotation (revs/unit time).
$\omega = 2\pi N$ angular velocity of rotation.
$U = \pi DN$ peripheral speed of journal.

ϕ = Attitude angle.
β = Angle from attitude line to film section considered.
h^* = Film thickness.
$e = \overline{OA}$ Eccentricity of journal.
μ = Viscosity of lubricant.
$n = \frac{e}{r}$ Eccentricity ratio.
$R = \frac{D}{2}$ Journal radius.
$r = \frac{c}{2}$ Radial clearance.

$$\text{Duty Parameter } S = \frac{\mu N}{PW}\left(\frac{D}{C}\right)^2 = \frac{\mu DLN}{W}\left(\frac{D}{C}\right)^2$$

FIG. 4.1. Bearing diagram.

figure 4.2, illustrate the analysis and define a number of the symbols used. Reference to the definition of viscosity is also made in Appendix V.

Figure 4.1 represents a full circular bearing of length L. The bearing centre is O and the journal centre is displaced to A; the attitude line through OA intersects the clearance between bearing and journal at its maximum and minimum values. The pressure-bearing film in the clearance space starts at angle β_1 and finishes at angle β_2.

The clearance space outside the pressure-bearing film, from β_2 to β_1, contains a film which is aerated or broken. The angles β_1 and β_2 may vary along the axial length of the bearing.

The same diagram shows the axes of reference for journal displacements and for forces transmitted from bearing to journal; Ox, Oy horizontal and vertical and Ok, Ol along and normal to the attitude line.

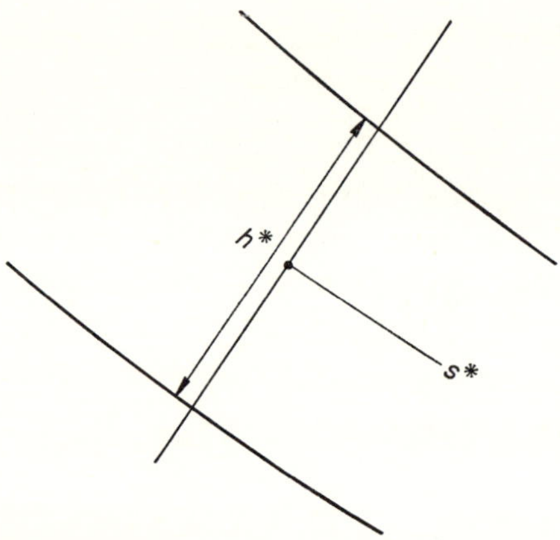

h^* = Film thickness.
s^* = Tangential dimension.
z^* = Axial dimension.
p^* = Pressure in film (relatively to atmospheric pressure).

FIG. 4.2. Diagram of oil–film section.

Some of the symbols are redundant; for example, journal diameter D and diametral clearance C are used as well as journal radius R and radial clearance r. Both sets of symbols are used, because the diametral dimensions are more convenient in engineering applications, while the radial dimensions are more convenient in analytical work.

It is shown in [9] that in a lubricating film for which the conventional assumptions hold good, the pressure distribution in steady running is determined by the basic differential equation

$$\frac{\partial}{\partial s^*}\left(\frac{\rho h^{*3}}{\mu}\cdot\frac{\partial p^*}{\partial s^*}\right)+\frac{\partial}{\partial z^*}\left(\frac{\rho h^{*3}}{\mu}\cdot\frac{\partial p^*}{\partial z^*}\right)=6U\frac{\partial(\rho h^*)}{\partial s^*} \tag{4.1}$$

Equation (4.1) remains applicable with non-circular bore of the bearing, and (subject to certain restrictions) with varying density and viscosity of the lubricant.

With constant density and viscosity, the factor ρ may be dropped, and μ may be taken outside the differentials.

To bring the equation to dimensionless form, put $\partial s^* = R\delta\beta$ and $U = 2\pi NR$, and introduce the dimensionless variables h, z, p, where

$$h = h^*/r,\ z = z^*/L,\ p = \frac{p^*}{N}\left(\frac{r}{R}\right)^2.$$

The expression for dimensionless pressure is obtained by inspection of the equation after the other variables have been made dimensionless.

The differential equation (4.1) becomes in dimensionless form

$$\frac{\partial}{\partial\beta}\left(h^3\frac{\partial p}{\partial\beta}\right) + \left(\frac{R}{L}\right)^2\frac{\partial}{\partial z}\left(h^3\frac{\partial p}{\partial z}\right) = 12\pi\left(\frac{\partial h}{\partial\beta}\right). \tag{4.2}$$

For a bearing of circular bore

$$h = 1 + n\cos\beta$$

and the differential equation is

$$\frac{\partial}{\partial\beta}\left\{(1 + n\cos\beta)^3\frac{\partial p}{\partial\beta}\right\} + \left(\frac{R}{L}\right)^2\frac{\partial}{\partial z}\left\{(1 + n\cos\beta)^3\frac{\partial p}{\partial z}\right\} = -12\pi n\sin\beta. \tag{4.3}$$

The conventional boundary condition for the pressure-bearing film is $p = 0$, with p measured relative to atmospheric pressure. The position of the boundary where this condition holds is known when the film starts at an oil-supply groove and when it ends at the side of the bearing or at an oil-drain groove. An additional assumption is required to determine the position at which the pressure-bearing film starts or terminates when this is not fixed by a supply or drain groove.

A widely used assumption (as Cameron and Wood [10]) takes the pressure-bearing film as starting at $\beta = 0$, the position of maximum clearance. This assumption is convenient for calculation, but is not fulfilled in practice unless there is sufficient oil available to form a continuous film at that position. Calculation shows, however, that if the film is assumed to start at a small angle after the position of maximum clearance the overall characteristics of the bearings are only slightly affected at normal operating attitudes.

The additional condition taken in conventional calculation as defining the position at which the pressure-bearing film terminates within the clearance is $\partial p/\partial\beta = 0$.

This condition is justified by the following argument which applies when $p = 0$ holds at the boundary – if $\partial p/\partial\beta$ had been positive there could have been a longer film with positive pressure; if $\partial p/\partial\beta$ had been negative there would have been negative pressure in the film.

The differential equation (4.3) applies only to a bearing of circular bore. It is applicable whether or not supply and drain grooves are present in the bearing surface. These grooves affect the boundary conditions. A supply groove defines the position at which a pressure-bearing film starts if the clearance is not divergent, while a drain groove defines the position beyond which a pressure-bearing film cannot extend.

Inspection of (4.3) shows that for the full circular bearing, solution of the equation with the specified boundary conditions is a function of two parameters, the width ratio $L/D = L/2R$ and the eccentricity ratio n. The first of these is determined by the bearing design and the second by the journal attitude. For given bearing design and chosen value of n, integration of the equation gives the nondimensional pressure distribution.

Under the load W the mean pressure on projected areas of journal is

$$p' = \frac{W}{DL} = \frac{W}{2RL}.$$

The dimensionless mean pressure, specified relatively to p' in the same way as the dimensionless local value of p is specified relatively to p^*, is

$$p_m = \frac{p'}{\mu N}\left(\frac{r}{R}\right)^2 = \frac{W}{\mu DLN}\left(\frac{D}{C}\right)^2.$$

The reciprocal of this, namely

$$S = \frac{\mu DLN}{W}\left(\frac{D}{C}\right)^2 \tag{4.4}$$

is the duty parameter as defined in a form convenient for engineering application. This duty parameter is nondimensional; any consistent set of units may be used to calculate it. For a bearing of specified design, operating in accordance with the assumptions of the conventional theory, the duty parameter determines the eccentricity ratio and the nondimensional operating conditions.

When the nondimensional pressure distribution has been obtained the values of p_m and ϕ are calculated from the component reactions on the journal. The dimensionless mean pressures of these component reactions, in the attitude axes k and l as shown in figure 4.1, are

$$\left.\begin{aligned} p_k &= \tfrac{1}{2}\int\int p \cos\beta \, d\beta dz \\ p_l &= \tfrac{1}{2}\int\int p \sin\beta \, d\beta dz \end{aligned}\right\} \quad (4.5)$$

where the integration extends over the area of the pressure-bearing film. The coefficient $\frac{1}{2}$ appears in these relations because the 'dimensionless projected area' of the journal, relatively to the unit LR, is 2.

The resultant p_m of these loadings must be vertically upward in equilibrium with the steady loading downward on the journal. For the axes here chosen p_k is negative.

Then $p_m^2 = p_k^2 + p_l^2$, which determines S, and ϕ is given by $\tan\phi = -p_l/p_k$.

In calculating the characteristics of a given bearing design, calculations of this type are carried out for a series of values of eccentricity ratio n. Additional calculations are then made which give power loss coefficient j and oil flow coefficient q as functions of n and hence of S. These dimensionless coefficients are defined in the following relations:

Calculated power loss $$H = jH_p \quad (4.6)$$

where Petroff loss $$H_p = 2\pi\mu LU^2\left(\frac{D}{C}\right)$$

$$= 2\pi^3\mu LD^2N^2\left(\frac{D}{C}\right) \quad (4.7)$$

Calculated oil flow $$Q = qQ_R \quad (4.8)$$

where Reference oil flow $$Q_R = \tfrac{1}{4}ULC. \quad (4.9)$$

The conventional power-loss calculation takes into account both viscous loss in the pressure-bearing film and viscous loss arising from the broken film in the clearance space. It does not include loss in oil grooves. Wilcock [8] assumes that the broken film is in viscous shear over a width corresponding to the oil flow in the film. The Wilcock assumption is used for the loss calculation here quoted. The calculated bearing loss is expressed nondimensionally relatively to the Petroff loss, which is the calculated viscous loss for the journal running centred in a complete bearing with unbroken film.

Analytical integration of equation (4.3) has been carried out only for the limiting cases of a very wide bearing and a very narrow bearing. In the case of the very wide bearing the conditions are taken as uniform axially, and the second term can be dropped from the left-hand side of the differential equation. In the case of the very narrow bearing an approximation due to Ocvirk [11] is used. It is assumed that for the narrow bearing the influence on the oil flow of the circumferential pressure gradient is everywhere of smaller order than that of the axial pressure gradient. This enables the first term to be dropped from the left-hand side of the differential equation.

For the fairly narrow bearings, often used in current practice, the Ocvirk calculation gives a good approximation to overall characteristics over a certain range of attitudes. It is, however, less accurate at high eccentricities, where the influence of the circumferential pressure gradient is not small compared with that of the axial pressure gradient. For a wide bearing, the angle at which the film ends is calculable analytically. For a narrow bearing, the Ocvirk assumption has the consequence that the film terminates at minimum film thickness, with $\beta_2 = \pi$.

When the Ocvirk approximation is used the parameter which determines the operating condition of the bearing is

$$S\left(\frac{L}{D}\right)^2 = \frac{\mu DLN}{W}\left(\frac{L}{C}\right)^2. \tag{4.10}$$

The results of the analytical calculations, so far as load–attitude relations are concerned, are given in Appendix I for the very narrow bearing and in Appendix II for the very wide bearing. For routine calculations curves drawn from the results are more serviceable than the formulae.

Analytical integration of the differential equation has not been accomplished for a bearing of finite width. Numerical solutions for steady running conditions in bearings of various designs have been established by various methods of computation.

In numerical calculation for a full bearing of finite width the position of the end boundary satisfying $p = 0$, $\partial p/\partial\beta = 0$ is found by a trial-and-error process. In this case the angle β_2 at which the film ends varies across the width of the bearing.

The oil flow is calculated relatively to a reference oil flow, which is the steady flow through a radial section of the clearance with journal centred and full film. The conventional oil-flow calculation, as carried out by Wilcock [8] assesses the oil escaping from the ends of the bearing.

It is assumed that oil passing into a supply groove is recirculated into the film formed after the supply groove. The oil-flow calculation is made for oil supply at atmospheric pressure, with the conventional assumption that the same pressure is maintained all along the oil-film boundaries.

In a bearing of circular bore which is divided by grooves into two or more bearing arcs calculation of characteristics can be made by the conventional method for each bearing arc with any specified journal attitude. The results of such calculations for certain partial arcs have been collected and tabulated by Wilcock [12]. For any attitude at which the clearance along the bearing arc is divergent, or is divergent–convergent with minimum clearance at entry, no pressure-bearing film is formed within that arc; the bearing arc then does not affect the calculated bearing load or the oil flow, but makes a contribution to the calculated bearing loss. The overall bearing performance at a given attitude is determined by calculation of the resultant load and the total oil flow and bearing loss for all the bearing arcs.

In the same reference Wilcock discusses the assessment of characteristics for non-circular bearings composed of circular arcs arranged eccentrically to one another. The same methods are applicable to calculation of the contribution of each bearing arc. In bearings of this type the number of independent design variables is increased, and geometrical complications affect the expression of the overall characteristics. Some results of a calculation of this type for the split elliptical bearing are given later.

Calculated characteristics of some journal bearings in laminar flow

Characteristics which have been calculated for certain bearing designs are reproduced and discussed.

(*a*) *Circular bearings with* $L/D = 1$

The results of hydrodynamic calculation of journal attitude for two bearing designs are shown in figure 4.3. Both bearings are of circular bore and of width ratio $L/D = 1$.

The characteristics of the full bearing are based on the calculations by Cameron and Wood [10]. The oil film is assumed to start at the position of maximum clearance. The Cameron and Wood calculations are widely regarded as the first satisfactory calculations for laminar flow in full journal bearings of finite width, and are frequently quoted in bearing literature.

The split cylindrical design (as Wilcock and Booser [8]) has bearing surfaces consisting of top and bottom arcs of 150° each, separated by 30° grooves at each side. Oil is supplied to each groove. The load is applied centrally. This bearing arrangement is representative of a

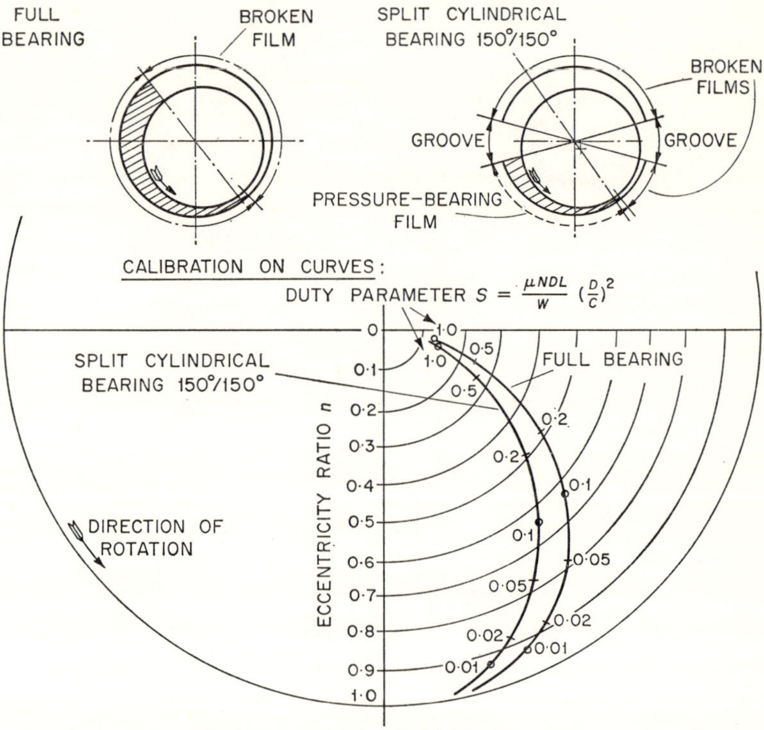

FIG. 4.3. Journal bearings: calculated attitudes circular. $L/D = 1$.

design used in turbomachinery. The starting-point of the pressure-bearing film is fixed by the groove on the descending side. At most journal attitudes the pressure-bearing film ends shortly after the point of minimum film thickness and before the groove on the other side is reached, as shown in the right-hand diagram of figure 4.3. At low eccentricity the film may reach the groove on the ascending side. The top bearing surface contains a broken film.

Values of the duty parameter S are marked along the attitude curves of figure 4.3. The two attitude curves are of similar shape, and for the same value of S the eccentricity ratio, and hence the minimum film thickness, is nearly the same for both bearings. The attitude curves illustrate the course of events when a turbine bearing under constant

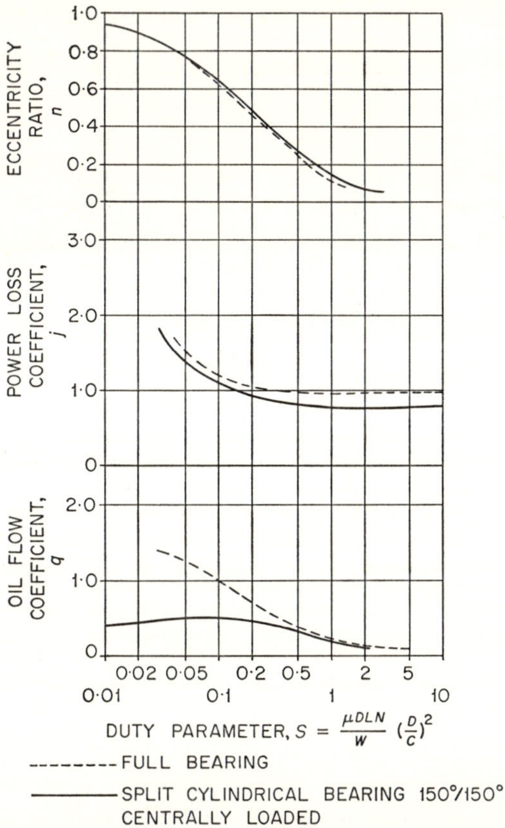

FIG. 4.4. Journal bearings: calculated characteristics. All curves for circular bore with $L/D = 1$.

load is run up from rest, without jacking oil, to normal operating speed. The journal is initially lying at the bottom of the clearance space. The first movement is made under conditions of boundary lubrication, but a hydrodynamic-bearing film forms at a low speed. As the speed rises, the journal climbs along the attitude curve to its normal running position.

Eccentricity ratio, power loss coefficient, and oil flow coefficient are plotted against S in figure 4.4; S is represented on a log scale. The slow variation of eccentricity ratio with S, as shown in figures 4.3 and 4.4, indicates the flexibility of a journal bearing in accommodating itself to a wide range of operating conditions, and its capacity for overload.

The power loss in the full bearing has been calculated using Wilcock's rule for loss in the broken film; this rule differs from that used by Cameron and Wood. For both bearings the power-loss coefficient is in the neighbourhood of unity for eccentricity ratios less than 0·7, but increases rapidly as the eccentricity ratio increases beyond 0·7, that is as the minimum film thickness becomes small.

The oil-flow coefficient is the only one in which there is a substantial difference between the calculated coefficients for the two designs of bearing. The 360° bearing has a higher oil-flow coefficient than the split cylindrical bearing, and the difference is particularly marked at high eccentricities. The explanation of the difference is that the split cylindrical bearing has a shorter pressure-bearing film, and so has less leakage from the ends of the bearing.

This comparison illustrates a general characteristic of journal-bearing designs, that provided there is a continuous bearing arc of substantial length in the bottom half, change of the grooving arrangement in the top half and sides has a limited effect on minimum film thickness and on loss coefficient, but may have great effect on the calculated oil flow.

The minimum film thickness in a circular bearing is

$$h_{\text{min}} = (1 - n)\frac{C}{2}. \tag{4.11}$$

(b) Split cylindrical bearing of varying L/D

In practice, a considerable range of values of L/D is used in bearings. Calculated characteristics of split cylindrical bearings with values of L/D from 0·25 to 1·5 are given in figure 4.5. These are replotted from results given by Wilcock and Booser [8]. The attitude curves for the various values of L/D are somewhat similar to one another in shape, but as the L/D ratio is decreased (that is as the bearing is made narrower) the value of S is increased for the same eccentricity ratio. Hence for a narrow bearing to run with the same minimum film thickness as a wider bearing of the same diameter, clearance, viscosity, and speed, the narrow bearing must have lighter loading on projected area.

The approximate Ocvirk theory for narrow bearings suggests that eccentricity ratio should be a function of $S(L/D)^2$ for bearings of varying

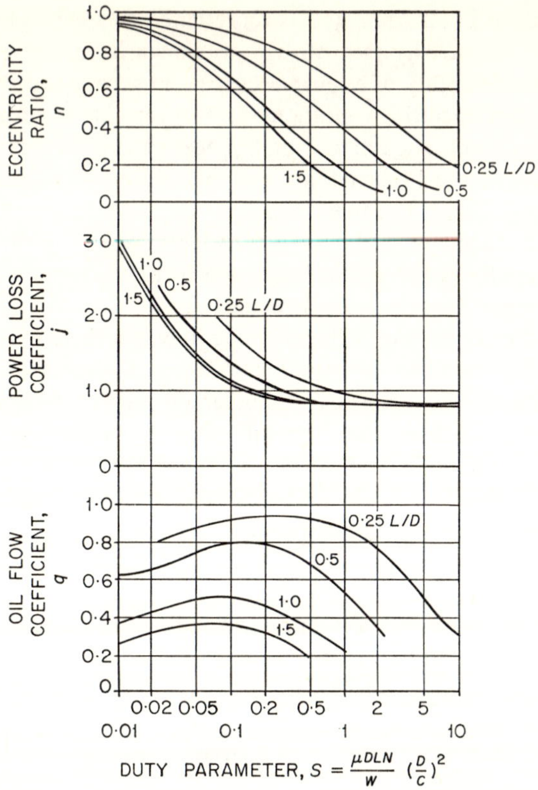

FIG. 4.5. Calculated characteristics of split cylindrical bearing. Split Cylindrical Bearing 150°/150° Centrally loaded.

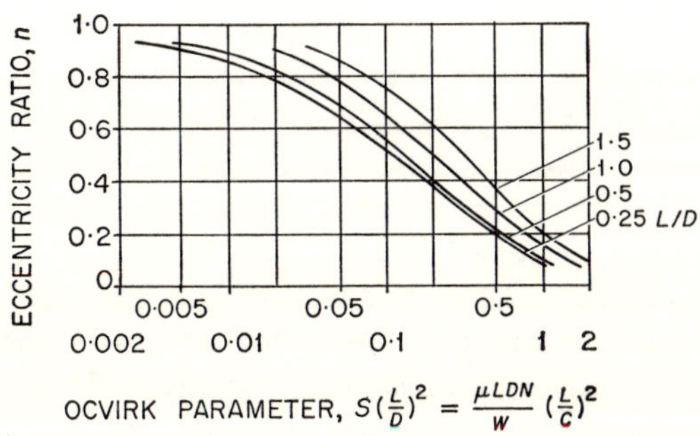

FIG. 4.6. Calculated characteristics of split cylindrical bearing. Eccentricity ratio as function of Ocvirk parameter.

(L/D). Figure 4.6 has been plotted to show the eccentricity ratios of figure 4.5 as functions of the Ocvirk parameter $S(L/D)^2$. It is seen that, plotted in this way, the eccentricity ratio is almost independent of L/D up to $L/D = 0{\cdot}5$, but there is departure from the Ocvirk relation for bearings of greater width. The curve shown on the figure for $L/D = 0{\cdot}25$ is close to that calculated using the Ocvirk assumption for a plain circular bearing, over the range of eccentricity ratios up to 0·8.

Returning to figure 4.5, for all values of L/D the power-loss coefficient is in the neighbourhood of unity when eccentricity ratio is below 0·7, but it rises at higher eccentricity ratios.

The oil-flow coefficient is higher the lower the value of L/D. This is to be expected, since as L/D falls, the reference oil flow falls in proportion, while the drain area at the ends of the bearing remains unchanged.

(*c*) *Split elliptical bearings*

Bearings of the type conventionally called elliptical are extensively used in turbine practice. These elliptical bearings are bored circular with shims on the horizontal joint, and the shims are removed before assembly, so that the vertical clearance in the assembled bearing is less than the horizontal clearance.

The dimensionless parameters S, n, j, and q have been defined for bearings of circular bore. Corresponding parameters can be defined for bearings which are not of circular bore, provided that the choice of clearance C is specified. In the case of the split elliptical bearing the most convenient choice for C is the clearance with which the bearing was originally bored. This is equal to the maximum clearance when the bearing is assembled (the horizontal clearance), and is also the diametral clearance corresponding to twice the difference in radius of curvature between each bearing land and the journal.

The formula (4.11) for minimum film thickness does not hold good for a non-circular bearing. The journal attitude is taken as specified by eccentricity ratio and attitude angle relatively to the bearing centre. To examine for minimum film thickness, an attitude diagram representing clearance in each transverse direction must be drawn for the bearing.

Figure 4.7 shows a bearing diagram and the corresponding attitude diagram for the split elliptical bearing. The bearing centre is o; the bottom and top land centres are o_1, o_2; the journal centre is o'. The definitions of the various eccentricity ratios are shown in the figure.

Calculations have been made by Pinkus [9] of the characteristics of split elliptical bearings with top and bottom lands of 150° arc. The limiting case with zero ellipticity is the Wilcock split cylindrical bearing. In most cases shown on the curves a pressure-bearing film forms on the bottom land only, but in some cases there are pressure-bearing films on both lands.

The calculated attitude curves for various values of the ellipticity

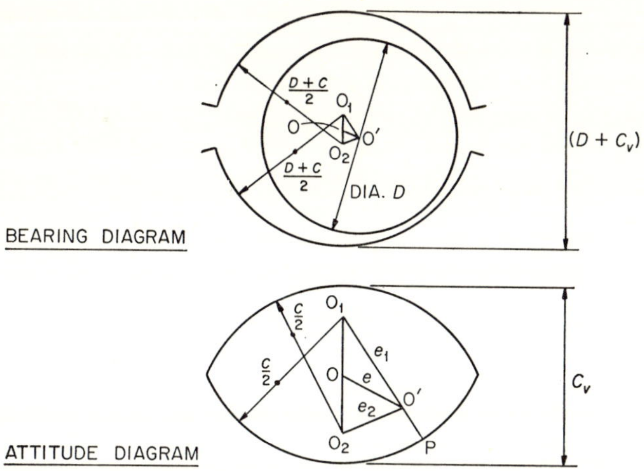

c = Horizontal clearance (Bore).
c_v = Vertical clearance.
$h_{\min}$ = O′P minimum film thickness.
$m = 1 - \dfrac{c_v}{c}$ ellipticity ratio.
$n = \dfrac{2e}{c}$ eccentricity ratio.
$n_v = \dfrac{2e}{c_v}$ eccentricity ratio referred to vertical clearance.
$n_1 = \dfrac{2e}{c}$ eccentricity ratio referred to bottom land.

FIG. 4.7. Split elliptical bearing.

ratio m are shown in figure 4.8. The curves are superimposed in two different ways – in the upper figure in terms of eccentricity ratio referred to vertical clearance (i.e. as for a series of bearings with the same vertical clearance and different horizontal clearance), and in the lower figure in terms of eccentricity ratio referred to bottom land (i.e. as for a series of bearings with different vertical clearance and the same horizontal clearance).

It will be noticed that for certain values of ellipticity ratio the attitude curve is calculated as rising above the bearing centre line. This happens only with high values of S, corresponding to lightly loaded bearings. The steady running attitude is unstable on that portion of the attitude curve which falls with rising S.

Although the journal may run above the bearing centre, it does not run above the centre of curvature of the bottom bearing land, when the load is applied vertically downwards.

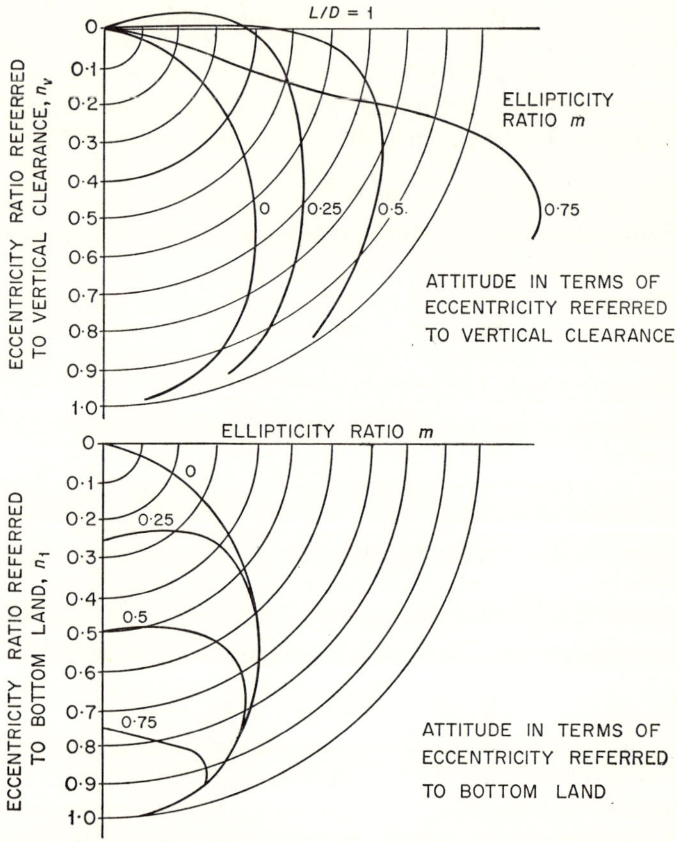

FIG. 4.8. Calculated characteristics of split elliptical bearings.

Calculations of the other bearing characteristics show that in bearings for the same service, with common horizontal clearance and varying vertical clearance, the power-loss coefficient increases with falling vertical clearance (i.e. with increasing ellipticity ratio). This increase is slow as the ellipticity ratio increases from 0 to 0·5, and is rapid at higher ellipticity ratios. The oil-flow coefficient varies little with ellipticity ratio over a wide range of conditions.

If comparison is to be made between bearing designs for the same vertical clearance and varying horizontal clearance the value of duty parameter changes with the horizontal clearance, and the characteristics must be separately calculated for each design.

Calculations of overall bearing characteristics for the range of bearing designs and operating attitudes normally used in turbomachinery show that the loss and the journal attitude relatively to bottom land are determined largely by the bottom land. The calculated oil flow, however, is greatly affected by the design of the grooves and the top lands.

(*d*) *Other bearing designs*

The journal bearings of fixed-land type used in turbomachinery generally have bearing lands which are circular arcs, although the complete bearing is often not of circular bore. As already mentioned, overall characteristics of these bearings can be built up from conventional calculations for the individual bearing lands. Some calculations of this type are given in refs. [8] and [9].

Tilting-pad journal bearings generally have pads with radius of curvature of the surface greater than the journal radius. This gives an effective pad clearance which can be used in applying the calculated characteristics of bearing arcs to establishing the characteristics of individual tilting-pads, for which the resultant load must pass through the pivot point. The calculated overall characteristics of a tilting-pad journal bearing may then be obtained by combining the characteristics of the individual pads. Allowance has, of course, to be made for the bearing clearance provided by the pivot setting; this is not the same as the clearance corresponding to difference in curvature.

In a tilting-pad journal bearing where the pad arrangement is symmetrical about the load line and the tilting action corresponds to free pivoting, the corresponding pad reactions are similar on opposite sides of the load line, and the calculated attitude line is coincident with the load line. There is no transverse displacement of the journal centre with load.

(*e*) *Calculation of effect of varying oil-supply pressure*

In ref. [8] Wilcock and Booser describe a method of calculating the effect on the oil flow of varying the oil-supply pressure. This consists essentially of adding the 'hydrodynamic flow' (the flow with atmospheric supply pressure as calculated for the bearing in operation) and the 'zero-speed flow' (the flow under the pressure difference with the journal at rest and centred in the bearing). These flows are assumed independent of one another.

The oil-flow variation with supply pressure is in fact influenced by the journal attitude, and the supply pressure has some effect on the attitude. The conventional assumptions could be applied to calculate these effects; but these calculations are complicated and are not carried out in practice. In turbine journal bearings the oil-supply pressure is much lower than the bearing load on projected area. The assumption that the correction for 'zero-speed flow' applies also under speed and load is a serviceable approximation for normal operating conditions. The actual zero-speed flow, with oil supplied under pressure and the loaded journal at rest in the bottom of the bearing, is much greater than the 'zero-speed flow' of the conventional calculation.

CHAPTER FIVE

Similarity Rules, Flow Regimes, and Characteristic Relations

Sommerfeld similarity

The hydrodynamic theory discussed in the previous chapter leads to the conclusion that with the conventional assumptions for laminar flow the characteristics in steady running of a journal bearing of specified design can be expressed non-dimensionally as functions of a single parameter – the duty parameter. 'Specified design' in this context does not imply geometric similarity. It requires, for a circular bearing, no more than geometrically similar boundaries of the bearing surface. For a non-circular bearing it is necessary to specify in addition some criterion defining proportionate circumferential variation of clearance; for example, the ellipticity ratio of an 'elliptical' bearing. The profile of the grooves, and the numerical value of the clearance ratio, do not enter into the specified design. The clearance ratio and the operating conditions are covered by the duty parameter.

This similarity condition, involving duty parameter alone, was first established for a journal bearing by Sommerfeld and is described as Sommerfeld similarity. Although the conventional assumptions are not fulfilled in practice, the duty parameter has given a serviceable measure of similarity conditions for many bearings with laminar flow in the film.

Reynolds similarity

Hydrodynamic similarity of flow in two different systems, each having passages filled with fluid of constant density and viscosity, is obtained when there is geometric similarity, similarity of initial flow distribution, and continued dynamic similarity between the two systems (ref. [13], section 4.7). Dynamic similarity requires that the forces, including inertia forces, acting on corresponding particles of fluid in the two systems should be similar. In the present context the only forces to be considered as acting on a particle are those arising from pressure difference and from viscous shear, giving as resultant the force producing acceleration of the particle. Dynamic similarity is obtained when the ratio of inertia force to viscous shear force is the same for corresponding small volumes of fluid in the two systems. This is equivalent to the dimensionless characteristic $(Re) = \rho V l_c / \mu$ having the same value in both systems, where V is a representative velocity and l_c is a representative length in correspondence in the two systems. (Re) is defined as the Reynolds number. When the Reynolds number is the same in both systems similarity of flow is ensured.

The conditions for similarity of flow in systems with fixed walls apply also to certain systems with walls in steady relative motion; for example, to a system including a journal rotating in a casing filled with fluid, provided the peripheral velocities of the journals in the systems compared give the same Reynolds number. The similarity laws do not tell what the flow is in any particular case; they state only that the flows are similar when all the conditions for similarity hold good. The nature of the flow, and the flow variation with Reynolds number and with geometry of system, must in general be established experimentally. In some cases the flow can be established by hydrodynamic analysis, but such calculation must be treated with caution in the absence of experimental verification, because a flow consistent with the hydrodynamic equations of motion is not necessarily stable.

Experience and analysis have shown that certain general statements about flow regimes can be made. When the Reynolds number in any system is sufficiently low the viscous forces predominate over the inertia forces, and at very low Reynolds numbers the flow approximates to that determined by the viscous forces in the absence of inertia. As the Reynolds number increases, the inertia forces increase relatively to the viscous forces so long as there is an approximately similar flow regime. A flow regime is, however, liable to undergo great change with change in Reynolds number.

For the journal bearing, the conditions outside the pressure-bearing film require consideration.

In the conventional bearing theory it is assumed that the pressure-bearing film is unable to sustain sub-atmospheric pressure, that its beginning, sides, and ends are at atmospheric pressure, that the journal surface is at atmospheric pressure outside the pressure-bearing film, and that sufficient oil is available to form a film wherever the geometry of the clearance space permits. These assumptions are not fulfilled in practice, but they provide an approximation to the conditions in many of the bearings to which the theory is applied.

For the purpose of establishing a 'conventional' similarity theory applicable to journal bearings for any flow regime in the film, the simplest procedure is to replace the assumption of negligible inertia of oil by taking the inertia of oil into account, and to abandon the assumption that there is laminar flow in the film, while retaining the other assumptions of the conventional theory. The broken film outside the pressure-bearing region then does not influence the load–attitude relation, although it contributes to the bearing loss. For a bearing of given design and operating conditions, this makes the extent of the pressure-bearing film and the journal attitude determinate, but not functions of duty parameter alone. These characteristics are also functions of Reynolds number. The word 'determinate' does not here imply that it is possible to calculate film extent and journal attitude by an available analytical method, but that these characteristics are physically determined by the imposed operating conditions.

A more comprehensive similarity theory can be adopted which takes the oil grooves into account. It is assumed that each supply groove is filled with oil which enters the groove at atmospheric pressure, and that each drain groove is either kept empty by drainage or is filled to a constant level at which it drains to atmospheric pressure. Geometric similarity of the supply and drain grooves must then be included in the specified similarity conditions. The consequence of including the grooves is that the similarity rule ceases to apply where there is a change in groove design. In practice, there are sometimes substantial losses in the grooves, and flow conditions in the grooves influence the conditions in the bearing-oil film.

In this more comprehensive similarity theory the condition of geometric similarity for the Reynolds criterion requires that all oil grooves which affect bearing flow shall be geometrically similar, that the bearing clearance ratio shall be the same, and that the journal attitude shall be the same. With geometrical similarity thus defined and with the same Reynolds number the flow conditions in two journal bearings are similar whatever the flow regime may be, provided the conventional assumptions retained hold good.

For definition of Reynolds number in a system it is possible to choose arbitrarily any characteristic length and any characteristic velocity in the system. It is convenient to choose a characteristic length and a characteristic velocity which have particular significance in the problem under investigation. This facilitates comparison between kindred systems of different geometry. For investigation of film flow in a journal bearing it is advantageous to choose the designed radial clearance as the characteristic length, since the local conditions across the oil film are of outstanding significance, and the designed clearance is independent of the attitude. The obvious choice for characteristic velocity of the journal is the peripheral speed of the journal. The oil density and viscosity are assumed constant in the conventional theory.

Reynolds number is defined as

$$(Re) = \frac{\rho}{\mu} U\left(\frac{C}{2}\right) = \frac{\pi}{2} \cdot \frac{\rho}{\mu} \cdot NDC. \tag{5.1}$$

Inertia of the oil has some influence on the film characteristics even while the film flow remains laminar; but the inertia effect in the film is usually small at speeds below that at which breakdown of laminar flow takes place. As previously mentioned, acceleration of oil at inlet to film may give a noticeable inertia effect. In a bearing with oil-filled grooves nonlaminar flow usually develops in the grooves before it develops in the film; this introduces an additional inertia effect associated with Reynolds number even when the film flow remains of laminar type.

J. A. Cole [14] has shown that in the equations of motion of the oil in the film, when written in a form which assumes laminar flow but takes inertia of the oil into account, the ratio of the magnitude of the inertia term to that of the predominant viscous term is of order $(Re)(C/D)$. This parameter determines the significance of the inertia of the oil in influencing the pressure distribution within the film.

Returning to the general case with any flow regime, the loads on similar bearings obey the Reynolds similarity rule for corresponding forces. The load relationship for conditions of Reynolds similarity in bearings can be written in several ways; for example,

$$W \propto UD \propto \mu ND^2 \propto \mu DLN. \tag{5.2}$$

Summarizing, when consideration is to be given to the possibility of various oil-flow regimes in the bearing film, the Sommerfeld similarity conditions are no longer sufficient. General hydrodynamic theory shows that if the conventional assumptions of the previous chapter are made, excluding those that inertia of oil is negligible and that flow regime in the film is laminar, similarity of flow is obtained when the following Reynolds conditions are fulfilled: flow passages geometrically similar (including geometrically similar grooving, the same clearance ratio, and the same journal attitude) and Reynolds number the same.

If bearings geometrically similar in design are run with duty parameter and Reynolds number both identical, then the journals run at the same attitude and the conditions for Reynolds similarity are fulfilled.

The conditions for Reynolds similarity in any flow regime are thus much more restricted than those for Sommerfeld similarity in the laminar-flow regime.

Flow between rotating cylinders

Published analytical and experimental investigations of flow in journal bearings deal mainly with the laminar-flow regime. The basic information on other flow regimes comes from hydrodynamicists who have investigated the stability of flow between rotating cylinders. This subject has received considerable attention.

The case primarily investigated by hydrodynamicists, both analytically and experimentally, is the simple case of flow in the annular space between long concentric cylinders, filled with a fluid whose viscosity and density can be treated as constant, with outer cylinder fixed and inner cylinder rotating. For these conditions G. I. Taylor [15] established in 1923 a criterion for instability of laminar flow.

The Taylor criterion for small clearance ratio may be expressed as

$$R_{\text{crit}} = 41{\cdot}1\left(\frac{D}{C}\right)^{\frac{1}{2}} \tag{5.3}$$

where R_{crit} is the Reynolds number of the oil film (as previously defined) above which laminar flow is unstable. R_{crit} corresponds to the condition

$$\frac{\rho}{\mu}ND^2\left(\frac{C}{D}\right)^{\frac{3}{2}} = 26{\cdot}2.$$

The Taylor criterion leads to the definition of the Taylor number:

$$N_T = (Re)\left(\frac{C}{D}\right)^{\frac{1}{2}} = \frac{\pi}{2}\frac{\rho}{\mu}ND^2\left(\frac{C}{D}\right)^{\frac{3}{2}} \tag{5.4}$$

which is a measure of liability to instability of the laminar flow regime. For the ideal case above mentioned the flow is stable when N_T is less than 26·2, and unstable when N_T is greater.

The Cole parameter $(Re)(C/D)$ is not identical in form with the Taylor number. Both parameters are dimensionless numbers which measure liability to inertia effects, but the inertia effects are of different types. The Cole parameter represents a two-dimensional effect on pressure distribution in steady flow in the film. The Taylor number represents liability to three-dimensional instability of flow in a curved clearance space.

The following information on flow regimes is based on various publications; refs. [14] to [19] are among them.

The tests in which direct observation has been made of nonlaminar flow regimes have been carried out with clearance ratios greater than those used in normal journal bearings. The Taylor criterion, when applied to operating conditions sufficiently close to the ideal conditions assumed for analysis, predicts the speed at which laminar flow becomes unstable for the range of clearance ratios tested. For the condition of outer cylinder fixed, inner cylinder in concentric rotation, and full film, it is consistent with the theory to extrapolate the observations of flow regime to clearance ratios of about $\frac{1}{1000}$ to $\frac{1}{2000}$ as used in turbine journal bearings. This gives the following sequence of flow regimes with rising speed to be expected for a centralized journal with full film in a circular bearing.

At the lower speeds the flow is viscous and of laminar type. At a speed corresponding to the Taylor criterion the laminar flow becomes unstable, and is replaced by a flow which is steady in space but consists of discrete vortices. This regime prevails over a certain speed range, but with further rise in speed the steady vortex motion is replaced by other vortex motions which are no longer steady in space. Continued rise in speed brings about a regime in which the discrete vortices are replaced by small-scale turbulence.

The Taylor criterion shows that increase of the film curvature, as represented by the clearance ratio (C/D), diminishes the Reynolds number for onset of instability. The onset of small-scale turbulence, however, is determined mainly by the local Reynolds number in the film, and at small clearance is little affected by the clearance ratio. At a sufficiently small clearance ratio the laminar regime would presumably pass into a regime of small-scale turbulence without an intermediate discrete vortex regime. It has been calculated that this would happen with clearance ratio near $\frac{1}{2500}$ [17]. This is a smaller clearance than is used in turbomachinery bearings.

For a bearing where the outer member rotates, whether the inner member rotates or not, the phenomena are greatly changed. This condition does not arise in current turbine practice.

Test experience shows that even small departures from the ideal conditions of a centralized journal in a long bearing have great effect on the onset of nonlaminar flow.

Analytical investigation has been made of stability of laminar flow in the case of long cylinders which are not concentric and which consequently set up a circumferential pressure gradient in the film [18]. The results show that where the pressure gradient is increasing in the direction of rotation there is an increased range of stability of the local laminar flow. Analysis for the case of axial-pressure gradient giving an axial-flow component shows that, for the case of concentric cylinders, axial-pressure gradient increases the range of stability.

In most of the analytical investigations of stability hydrodynamicists have used a linear theory to investigate the stability of small disturbances imposed on laminar flow. This linear theory does not give the fully established nonlaminar flow, and does not predict which motion will be set up when there is more than one possible mode of instability. Some progress has been made in application of nonlinear theory to these problems.

For the centralized journal with full film the observed loss in laminar flow is in approximate accordance with the Petroff relation. Beyond the laminar-flow region the loss is higher. At onset of discrete vortex flow the dimensionless loss coefficient increases rapidly with speed; this increase provides a convenient experimental indication of transition speed.

In ref. [19] Donnelly and Simon have examined torque characteristics obtained experimentally under laboratory conditions. In an appendix to this paper, G. K. Batchelor has analysed an 'inviscid core and viscous boundary layer' model for the discrete vortex-flow regime. His result corresponds to

$$\frac{\text{Loss in vortex flow regime}}{\text{Petroff loss}} \propto N_T^{\frac{1}{2}}$$

where N_T is the Taylor number. The test results analysed in the paper are approximately consistent with Batchelor's relation over a certain range of speed well above the transition speed.

Flow in journal bearings

The flow conditions in loaded journal bearings of the type used in turbomachinery are more complicated than the conditions for which the basic knowledge of flow regimes has been established. The bearings are of limited width; the bearing surfaces are not necessarily cylindrical and are interrupted by grooving for oil supply and drain; the lubricant is of varying viscosity in different parts of the film; under load the journal runs eccentrically, with a pressure-bearing film covering only part of the bearing surface, the remaining bearing surface being covered by a broken film. It is therefore not to be expected that the flow regime in turbine bearings can be defined by such clear-cut rules as apply to flow between long concentric rotating cylinders.

Observations of flow regime in loaded journal bearings have been made only for small bearings under laboratory conditions. Nonlaminar flow in loaded bearings has been observed by Cole [13] with large clearance ratios and lubricants of low viscosity. Discrete vortex flow was found of the type described by Taylor. At low eccentricity the transition speed was approximately in accordance with the Taylor criterion; increase of eccentricity raised the transition speed.

The experimental evidence of nonlaminar flow in journal bearings is provided largely by observations of power loss. The first significant series of tests, by Wilcock [20] (results summarized in [8]) covered only lightly loaded bearings running at low eccentricity; they showed transition near the Taylor criterion. A series of tests by Smith and Fuller [21], using water as lubricant, brought out some of the complications associated with bearing loading. Tests going well into the nonlaminar region were published by Neale and Love [22] on oil-lubricated bearings of 2 in. diameter. Two papers giving test results on large turbine bearings in nonlaminar flow were published in 1966 – by Huggins [23] and by Duffin and Johnson [24]. All the tests show substantial increase in loss coefficient as nonlaminar flow develops.

Some of the papers, notably ref. [23], include observations of journal attitude which provide independent evidence of the development of nonlaminar flow in the main bearing film. With nonlaminar flow the minimum film thickness is greater than it would have been at the same duty parameter with laminar flow. The eccentricity ratio is influenced by Reynolds number and is less influenced by duty parameter.

As judged by the criterion of minimum film thickness, the load-carrying capacity of a bearing is increased by the development of nonlaminar flow. The operating disadvantage occasioned by the nonlaminar flow in large turbine bearings is the high power loss. The oil-flow coefficient is little affected by transition to the nonlaminar flow regime.

Some writers on journal bearings use the word 'turbulent' to describe any flow regime which is nonlaminar, while others use the word only to describe small-scale turbulence; they do not apply it to discrete vortex flow. The use of the word nonlaminar avoids misunderstanding.

Flow in tilting-pad journal bearings

In a tilting-pad journal bearing in steady running the flow regime under the pads is the same as it would be under corresponding fixed pads with the same geometry, attitude, and oil-supply conditions; but in practice a tilting-pad journal bearing is not used with pads reproducing the geometry and attitude of the bearing surface of a plain journal bearing. In a tilting-pad journal bearing increasing speed gives a series of oil-flow regimes which have some correspondence with those in a plain journal bearing. The overall behaviour with changing speed, however, is different in the two bearing designs, because the tilting of the individual pads gives a sequence of pad attitudes different from the sequence of attitudes of the fixed bearing surfaces. The angle of tilt of the pad at given pivot clearance is affected by the flow regime.

The characteristics of a tilting-pad journal bearing running beyond the laminar region can be determined only by test. Some forecast of the performance of a new design of tilting-pad journal bearing can be made if test information is available of the characteristics of the individual pads.

Summary and discussion of characteristic relations

For convenience, certain formulae established in the previous chapter are collected below. These represent characteristic relations in steady running given by the conventional hydrodynamic theory of journal bearings with laminar flow in the film.

Duty parameter $$S = \frac{\mu DLN}{W}\left(\frac{D}{C}\right)^2 \qquad (4.4)$$

Eccentricity ratio $$n = \frac{2e}{C} \qquad (4.5)$$

Petroff loss $$H_p = 2\pi^3 \mu L D^2 N^2 \left(\frac{D}{C}\right) \qquad (4.7)$$

Calculated power loss $$H = jH_p \qquad (4.6)$$

Reference oil flow $$Q_R = \frac{1}{4}\, ULDC = \frac{\pi}{4} DLNC \qquad (4.9)$$

Calculated oil flow $$Q = qQ_R \qquad (4.8)$$

In a bearing which is not of circular bore the clearance C taken in calculation is twice the difference between radius of curvature of the bottom land and radius of journal; the eccentricity ratio is arbitrarily defined in some convenient way.

For a bearing of specified design (in accordance with the requirements of Sommerfeld similarity) the following nondimensional characteristics:

Eccentricity ratio	n
Attitude angle	ϕ
Power loss coefficient	j
Oil flow coefficient	q

are functions of the duty parameter S.

These relations have been calculated for various specified designs.

The Sommerfeld similarity conditions which hold good for the conventional hydrodynamic theory with laminar flow in the film do not ensure similarity when a nonlaminar flow regime develops in the film. In the present chapter a similarity theory is developed which covers any flow regime in the film. In this theory the density of the oil is assumed constant but not zero; the assumption of laminar flow is abandoned; and loss in the oil grooves, as well as in the film proper, is taken into account. The other assumptions of the conventional hydrodynamic theory of Chapter 4 are retained.

With this wider hydrodynamic theory it is found that for similarity in bearing operating conditions there must be geometric similarity of the bearings compared (including similar grooving and the same clearance ratio), and both duty parameter S and Reynolds number (Re) must be the same.

Reynolds number is defined as

$$(Re) = \frac{\rho}{\mu} U \frac{C}{2} = \frac{\pi}{2} \cdot \frac{\rho}{\mu} \cdot DNC. \tag{5.1}$$

The power loss is expressed by

$$H = K_j H_p = K_j \left\{ 2\pi^3 \mu L D^2 N^2 \left(\frac{D}{C} \right) \right\} \tag{5.5}$$

wher K_j is the power-loss coefficient.

The oil flow is expressed by

$$Q = K_q Q_R = K_q \left(\frac{\pi}{4} DLNC \right) \tag{5.6}$$

where K_q is the oil-flow coefficient.

For bearings of a given geometrically similar design, the non-dimensional characteristics n, ϕ, K_j, and K_q are functions of both S and (Re).

The relations can be written in various alternative forms. For some purposes it is convenient to express the loss in terms of the friction factor

$$f = K_j\left(\frac{2}{(Re)}\right) \tag{5.7}$$

giving

$$H = f\left(\frac{\pi^4}{2}\rho L D^4 N^3\right) \tag{5.8}$$

The friction factor is a function of bearing design, S and (Re).

Substitution gives $f = 4T/\pi\rho U^2 D^2 L$, where T is the torque $H/2\pi N$ which gives rise to the power loss. Hence f has the physical meaning:

$$f = \frac{\text{Mean fluid shear stress on journal surface}}{\frac{1}{2}\rho U^2}$$

It is of interest to note that

$$\frac{Q}{L} = K_q \cdot \frac{\mu}{\rho} \cdot \frac{(Re)}{2} \tag{5.9}$$

showing that high Reynolds number is associated with high oil flow.

The conventional hydrodynamic theory with laminar flow in the film permits of calculation of the characteristics of specific designs, as shown in the previous chapters. Similar calculation of characteristics from theory is impracticable for nonlaminar flow regimes. The analytical investigations in this field have included attempts to construct empirical laws of nonlaminar flow, giving consistency with particular series of experimental results. The analysis here presented is limited to consideration of certain approximate relations which provide guidance as to variation of characteristics with certain changes in design and operating conditions. These approximate relations are based on test results with service bearings, supplemented by certain laboratory tests on small bearings outside the service range, and with the test results analysed in the light of hydrodynamic theory. The only claim to validity is approximate consistency of the relations with these test results over a limited range.

Experience has shown that flow coefficient K_q is greatly affected by bearing design and by pressure of oil supply, but that over a considerable range, from a speed well within the laminar flow range to the highest speed tested, variations in speed and in load, within the range of normal operation, have relatively little effect on K_q. This flow coefficient may therefore be taken as independent of both (Re) and S for approximate calculation. K_q may, however, be substantially modified at operating conditions which give abnormal journal attitude.

Over a certain range of conditions near and above the speed of transition to nonlaminar flow the loss coefficient K_j appears to be influenced more by the Taylor number N_T than by other parameters.

Assume that over a certain range

$$K_j \propto N_T{}^m \tag{5.10}$$

where m is a constant index. Tests on large bearings suggest that over limited ranges of speed m varies between zero in the laminar-flow regime to a maximum which is slightly greater than unity in a portion of the discrete-vortex flow regime.

Then $$H \propto \rho^m \mu^{1-m} L D^{2+2m} N^{2+m} \left(\frac{D}{C}\right)^{1-\frac{3}{2}m} \tag{5.11}$$

When $$m = 0{\cdot}5,\ H \propto \rho^{0{\cdot}5} \mu^{-0{\cdot}5} L D^3 N^{2{\cdot}5} \left(\frac{D}{C}\right)^{0{\cdot}25}. \tag{5.12}$$

This formula is in agreement with Batchelor's analysis for his 'inviscid core and viscous boundary layer' model of discrete vortex flow.

When $m = 1$, $$H \propto \rho L D^4 N^3 \left(\frac{C}{D}\right)^{0{\cdot}25}. \tag{5.13}$$

In this expression loss is proportional to density and independent of viscosity, and loss increases with increasing clearance ratio.

Some journal bearings, in the highest speed range at which they have been tested, may have reached a flow regime of small-scale turbulence. It is to be expected that in this condition the loss coefficient will be predominantly a function of the Reynolds number (Re). Assume that over a certain range in this flow regime

$$K_j \propto (Re)^{m'} \tag{5.14}$$

where m' is a constant index.

Then $$H \propto \rho^{m'} \mu^{1-m'} L D^{2+2m'} N^{2+2m'} \left(\frac{D}{C}\right)^{1-m'} \tag{5.15}$$

For a large bearing mentioned in the next chapter the index m' was approximately 0·75 in the highest speed range tested.

With $m' = 0{\cdot}75$ the loss relation becomes

$$H \propto \rho^{0\cdot75}\mu^{0\cdot25}LD^{3\cdot5}N^{2\cdot75}\left(\frac{D}{C}\right)^{0\cdot25}. \tag{5.16}$$

The results for both the Taylor number and the Reynolds number hypotheses suggest that in the nonlaminar-flow regime loss is more affected by density of lubricant and less affected by viscosity of lubricant than in the laminar-flow regime, that loss is even more rapidly affected by journal diameter and speed, and that it is less affected by clearance ratio. It is possible that in certain cases loss may be increased and not diminished by increase of clearance ratio.

The variation of loss coefficient with Reynolds number experienced in journal bearings resembles the variation found in some other hydrodynamic systems of interest to engineers.

The maximum temperatures reached in the oil and in the bearing material have significant effects on the margin of safety in operation. Neither of these maxima can be readily measured or assessed on bearings in service because of uncertain local variations, but a useful temperature criterion for bearings of established design is provided by the mean oil-outlet temperature. Assessment of this temperature is required both as a criterion of safety and for determination of the effective viscosity to be taken in certain hydrodynamic calculations.

The oil-temperature rise is affected by heat conducted to the bearing and by heat radiated from the bearing, as well as by the power loss in the bearing. Where there is substantial heat exchange with the environment, this must be taken into account in calculating oil-temperature rise. In most bearings in turbomachinery, other than very small bearings, the net heat exchange with environment is small compared with the bearing loss, and may be neglected in approximate calculation. In the formulae given below the heat imparted to the circulating oil is assumed equal to the bearing loss.

The mean oil temperature rise is $t = t_2 - t_1$, where t_1 is oil-inlet temperature and t_2 is mean oil-outlet temperature. The power loss is related to this temperature rise by

$$H = J\rho c_p Q t \tag{5.17}$$

where J is the mechanical equivalent of heat and ρ and c_p are density and specific heat of the oil. For normal service conditions ρ and c_p can be treated as constant over the range of temperature rise.

Hence

$$t = \frac{K_j}{K_q} \cdot \frac{8\pi^2}{J\rho c_p} \cdot \mu N \left(\frac{D}{C}\right)^2. \tag{5.18}$$

The alternative relation in terms of friction factor is

$$t = \frac{f}{K_q} \cdot \frac{2\pi^3}{Jc_p} \cdot D^2N^2\left(\frac{D}{C}\right). \tag{5.19}$$

These relations illustrate the operating-temperature problems which arise at high speeds of rotation and with large-diameter bearings. For normal operation in the laminar-flow regime of a series of bearings designed with Sommerfeld similarity (K_j/K_q) is nearly constant. With given oil-supply conditions, and with constant clearancc ratio, the oil-outlet temperature is nearly independent of diameter and rises with speed. The actual temperature rise is not proportional to speed, because with lubricating oil the effective viscosity μ falls as the outlet-temperature rises. Temperature rise is diminished by increasing clearance ratio C/D. It is the practice of some turbine manufacturers to use approximately constant clearance ratio up to 3,600 r.p.m. and to use larger clearance ratios at higher speeds.

In the nonlaminar flow regime problems arise at 3,000 r.p.m. and above. K_j increases with increasing Taylor number. On large bearings considerations of journal location restrict the acceptable clearance ratio. Bearing design may have to be modified to give high flow coefficient K_q.

CHAPTER SIX

Observed Bearing Performance

Operating conditions in service

The operating conditions of a turbine journal bearing in service are not closely predictable, even when the design procedure ensures that the bearing will function satisfactorily and with high reliability. Uncertainty applies especially to the inter-related characteristics of oil flow, oil-temperature rise, and bearing loss, for which the designer has to allow considerable margins. The oil to be used in a lubricating-oil system is normally specified as being of a standard grade; but the standard specification defines only maximum and minimum viscosities at certain temperatures (Appendix VI).

When oil at atmospheric pressure is supplied to a bearing a flow is passed by the pumping action of the bearing. When oil is supplied at higher or lower pressure the oil flow is increased or diminished. The pumping action of the bearing is greatly influenced by the detailed arrangement of the oil-supply passages and grooves and by the bearing clearance, as previously mentioned. Imperfections in workmanship, which are acceptable from an operating point of view, and are within the specified tolerance limits, have considerable influence on the oil flow.

When the oil-supply pressure to a bearing is changed, with constant journal speed and load, there is a direct effect on bearing characteristics through the change in oil-supply pressure causing redistribution of the pressure in the bearing-oil film, as well as an indirect effect through the change in oil flow leading to a change in oil-temperature rise, and hence in viscosity variation through the bearing. In turbine bearings the indirect effect from viscosity variation usually outweighs the direct effect from pressure redistribution in its influence on the bearing operating characteristics.

The heat taken up by the circulating oil does not necessarily correspond to the bearing loss. Heat is lost by radiation from the bearing housing to its surroundings, while heat is often added by conduction along a turbine rotor to the journal. In rig testing of large turbine bearings, and on most bearings in service, it is permissible to take the bearing loss as represented by the heat input to the oil.

Performance observations on turbine bearings in service are nearly always limited to inlet-oil pressure and temperature and outlet-oil temperature. Systematic records of these observations are kept, since abnormal variation may give advance warning of trouble. Oil flow and bearing loss are rarely observed in service, but are sometimes measured in works test. Full-scale observations of journal-bearing performance over a range of operating conditions are made only on a few bearings in rig test. The bearings thus rig tested generally represent new developments in size or type.

As an introduction to the discussion of observed bearing characteristics, a series of test results will be given for a bearing which is representative of current practice on large turbines.

Test performance of 19 in. × 10 in. three-land bearing

The test results given in figures 6.1–6.6 were taken in rig test of a three-land bearing, 19 in. diameter × 10 in. in length, with circular bore of diametral clearance 0·023 in. Some of these test results have been previously published [24].

The test results are given for applied conditions which are standard or which vary one at a time from the standard conditions of oil-inlet pressure 10 lb/sq in., oil-inlet temperature 120° F, load on projected area 200 lb/sq in. corresponding to total bearing load of 17 tons. The results are given for various speeds. The design speed is 3,000 r.p.m.

Figure 6.1 shows the variation of power loss with speed from 750 to 3,000 r.p.m. With logarithmic plotting there is a rapid increase in gradient of the curve near 2,000 r.p.m. Two test points, with losses differing by 9%, are shown at 2,000 r.p.m. Some anomalies in test results were observed at this speed.

Figure 6.2 shows the variation of oil flow with oil-inlet pressure and with speed. The curve for 2,250 r.p.m. is included, since a wide range of oil-inlet pressure was applied at this speed. Over a substantial range of running conditions the oil-flow variation is nearly linear with speed and with oil-inlet pressure. The oil-flow variation with pressure is more rapid at zero speed (where the journal is resting at the bottom of the bearing) than at the running speeds. In the running condition the oil-flow variation is more rapid with low oil-inlet pressure than with moderate oil-inlet pressure.

Figure 6.3 shows the variation of power loss with load on bearing at certain speeds. The variations with load at each speed are of limited range. The bearing loss rises with load at 1,500 and 2,000 r.p.m., and falls slowly with load at 3,000 r.p.m. Oil-temperature rise with varying load passes through a minimum at each speed.

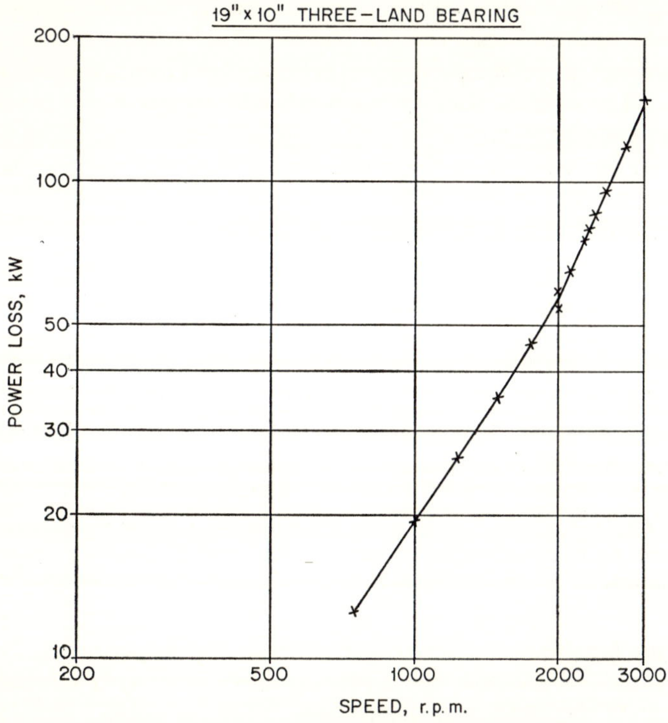

FIG. 6.1. Variation of power loss with speed.

The variation of power loss with oil-inlet temperature is shown on figure 6.4. At 1,500 r.p.m. the power loss falls with increase in oil-inlet temperature. At 3,000 r.p.m. the curve is of a curious shape, showing three sections of approximately constant loss, increasing loss, and constant loss with oil-inlet temperature increasing from 110° to 140° F. At 1,500 r.p.m. the oil-temperature rise through the bearing falls by about $\frac{1}{3}$ of the increase in oil-inlet temperature, while at 3,000 r.p.m.

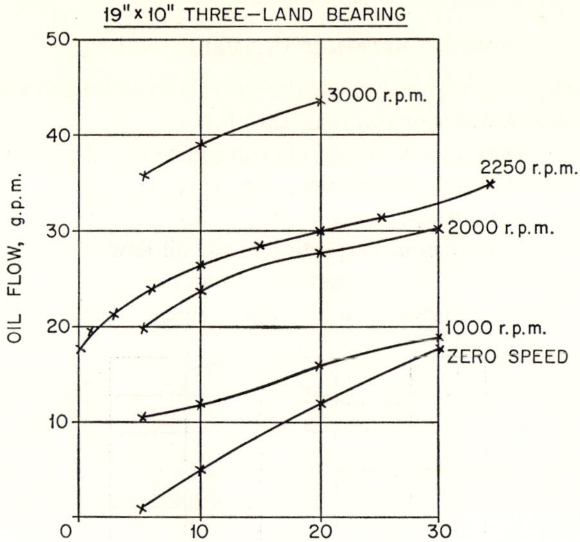

FIG. 6.2. Variation of oil flow.

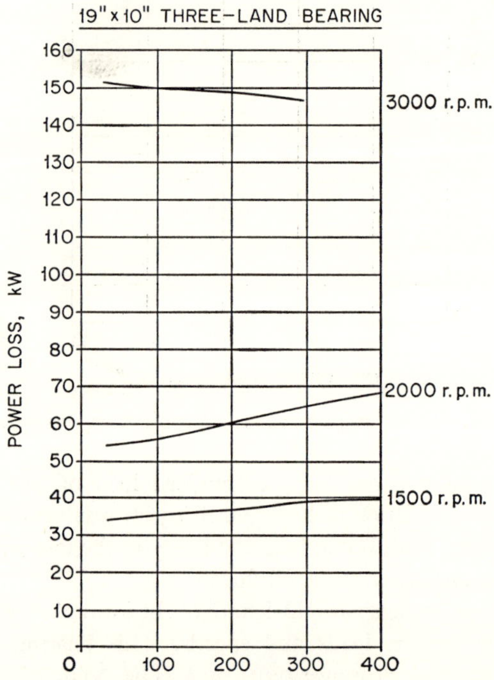

FIG. 6.3. Variation of power loss with bearing load.

the oil temperature rise through the bearing is about the same at the oil-inlet temperatures of 100° and 140° F.

Figure 6.5 shows the variation of power loss and oil-temperature rise with oil flow through the bearing. The speeds plotted are 3,000 and 2,250 r.p.m. At both speeds the loss rises substantially with oil flow, but much less rapidly than in proportion to the oil flow.

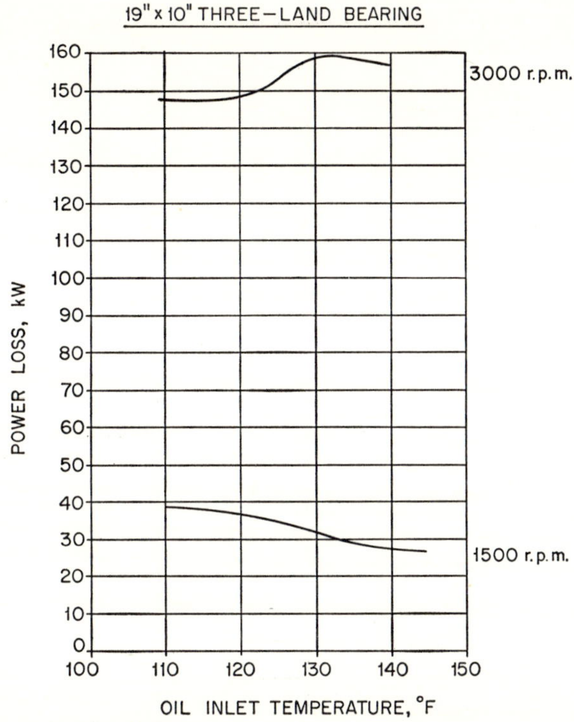

FIG. 6.4. Variation power loss with oil-inlet temperature.

The curves plotted are for the standard load of 200 lb/sq in. At 3,000 r.p.m. the variation at load 300 lb/sq in. was nearly the same as with the standard load, but at load 100 lb/sq in. the loss passed through a minimum at an intermediate oil flow.

Attitude measurements at mid-length of the journal axes are shown in figure 6.6 for various loads and speeds. This bearing had circular clearance (diametral clearance 0·019 in.). The points do not lie on a single locus, but are near to a single attitude line.

Figure 6.7 shows the viscosity variation with temperature for the lubricating oil used in the tests. The viscosity is plotted on a logarithmic scale. The rapid variation of viscosity with temperature is typical of turbine oils. The oil was to the British Standard Specification of 1955 for Medium Grade oil. A comparison of the specified range of kine-

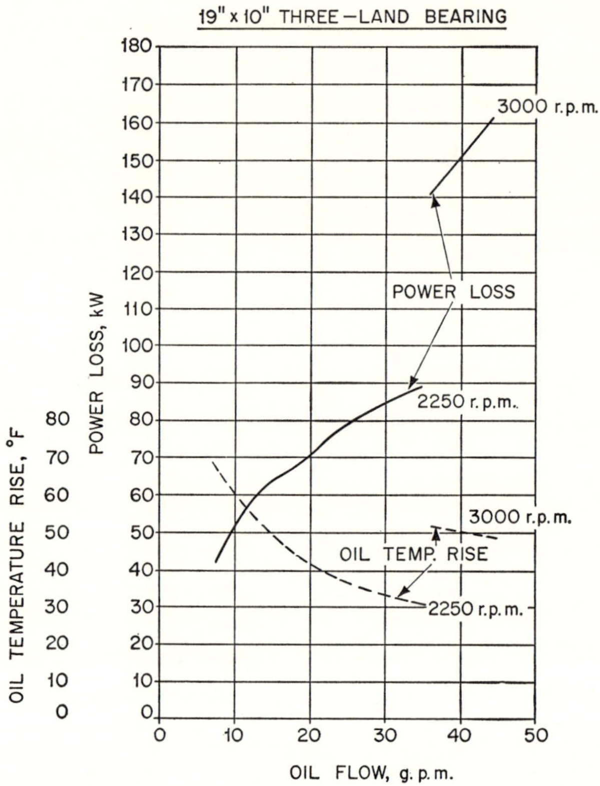

FIG. 6.5. Variation of power loss and oil-temperature rise with oil flow.

matic viscosity with the observed properties of the oil is given in Appendix VI.

The tests made on this bearing included measurements of oil-film pressure at a number of points on the bearing surface. An important observation is that sub-atmospheric pressures were measured at tapping points near the trailing edge of the bottom bearing land. At 3,000 r.p.m.

the extreme sub-atmospheric pressures observed were 3·5 lb/sq in. with load 200 lb/sq in. and 4 lb/sq in. with load 300 lb/sq in. It was found that oil-film pressures built up in the top lands of the bearings, but with load 200 lb/sq in. and above these pressures were small compared with the pressures in the bottom land.

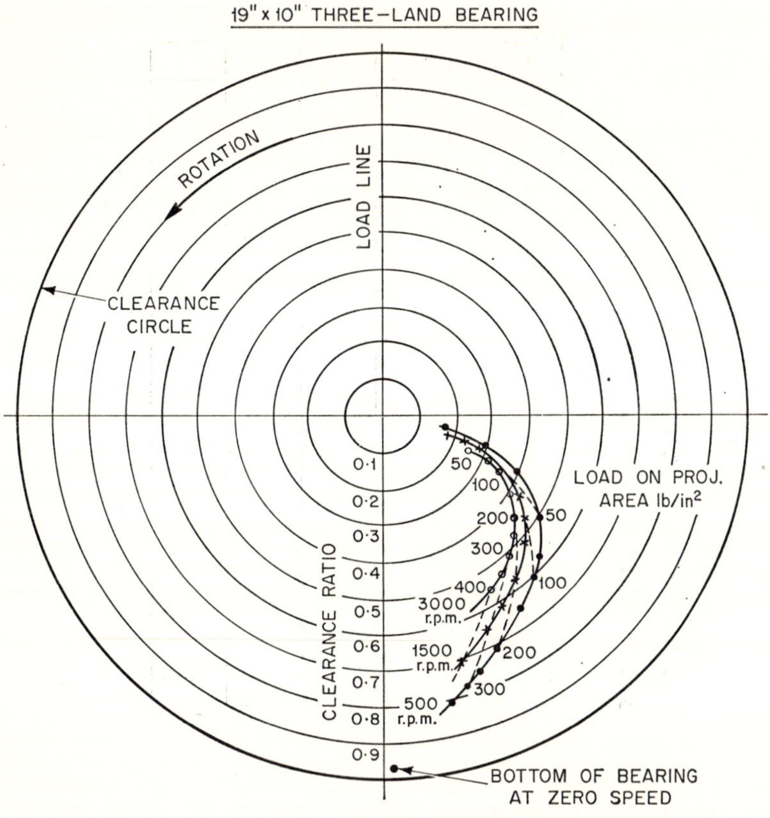

FIG. 6.6. Attitude variation with load and speed.

Observations of bearing-metal temperature were also made at a number of points. The maximum bearing-metal temperature was always much higher than the mean outlet oil temperature, the excess at 'standard' condition at 3,000 r.p.m. being about 60° F. With increasing load at 3,000 r.p.m. the excess increased to 67° F at 300 lb/sq in.

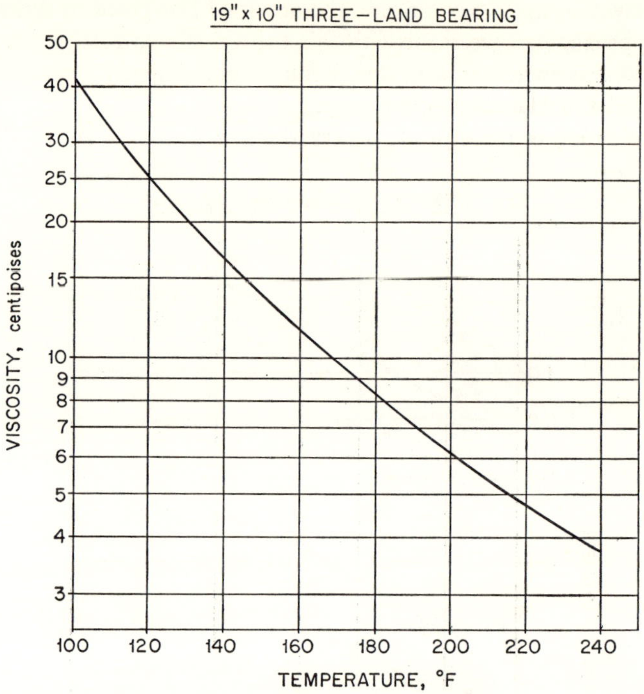

Fig. 6.7. Viscosity of oil used in tests.

Comparison of test performance with hydrodynamic theory

The two dimensionless parameters which play particularly significant parts in the hydrodynamic theory – duty parameter and Reynolds number – are plotted logarithmically in figures 6.8 and 6.9 respectively. These parameters are arbitrarily defined; the arbitrary definitions include choice of the oil viscosity as that at the mean oil-output temperature. Whether the assumptions of the hydrodynamic theory are reasonable, and whether this choice of effective oil viscosity is satisfactory for calculation of parameters, must be assessed by comparison of calculated results with observed bearing performance.

The Reynolds number is practically independent of load, while the duty parameter is inversely proportional to load. Owing to the effect of viscosity variation, the Reynolds number varies more rapidly and the duty parameter less rapidly than in proportion to speed.

Calculations of bearing loss from the conventional theory of laminar flow show that up to 2,000 r.p.m. the observed loss given by figure 6.1 is in approximate agreement with this theory. Above 2,000 r.p.m. the observed loss increases with speed much more rapidly than the loss calculated from laminar-flow theory.

Examination of the attitude observations of figure 6.6 shows that at

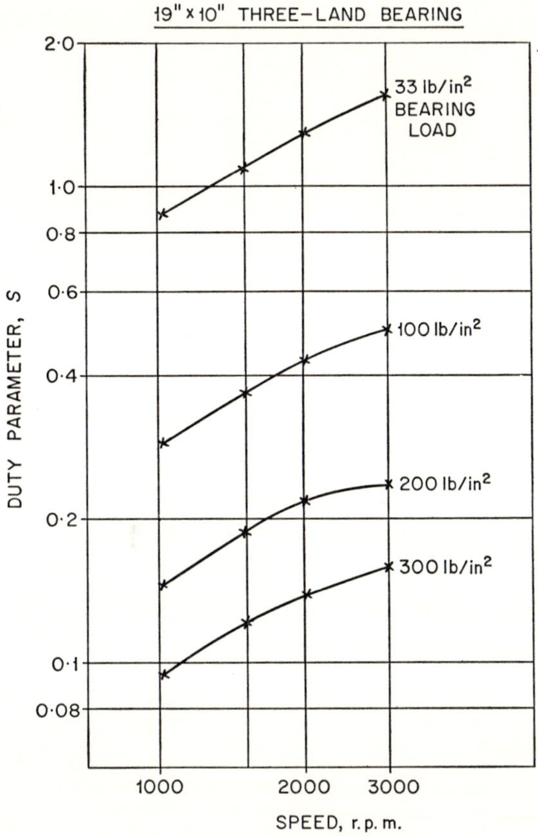

FIG. 6.8. Duty parameter.

500 and 1,500 r.p.m. the attitude loci approximate to the shape predicted by the conventional theory of laminar flow, and the attitude of points on each of these curves is nearly the same for the same duty parameter. At 3,000 r.p.m. the eccentricity ratio is lower for the same duty parameter, but the shape of the locus is not much changed.

The loss and attitude observations are reasonably consistent with the predictions of laminar-flow theory up to 2,000 r.p.m., assuming the effective viscosity as equal to the viscosity at oil-outlet temperature. The observed conditions diverge progressively from simple laminar theory at higher speeds.

For this bearing the critical Reynolds number given by the Taylor

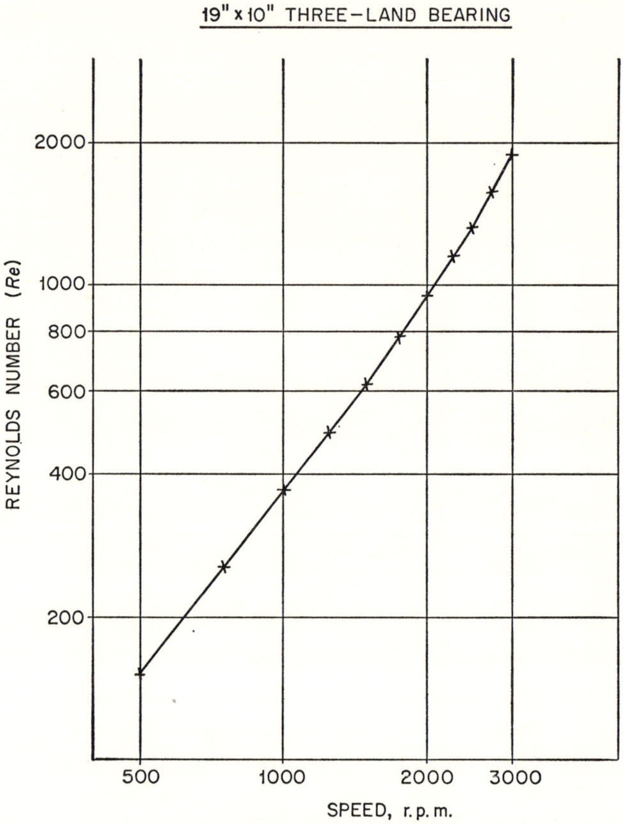

FIG. 6.9. Reynolds number.

criterion is 1,180, corresponding to the Reynolds number as conventionally defined at 2,250 r.p.m.

In figure 6.3 the slow increase in power loss with load at 2,000 r.p.m. is consistent with conventional laminar-flow theory, while the slow decrease with load at 3,000 r.p.m. is not.

Plotting of the friction factor, as defined in the previous chapter, against Reynolds number offers a convenient way of examining the power-loss characteristics of various bearing designs. Results thus plotted in logarithmic coordinates are shown in figure 6.10 for two builds of the 19-in. × 10-in. bearing, with diametral clearances 0·019 and 0·023 in. respectively, and for one build of a 24-in. × 18-in. elliptical bearing with clearances 0·034 in. horizontal, 0·026 in. vertical. The line representing $f = 2/(Re)$, corresponding to Petroff losses in a circular bearing, is also shown.

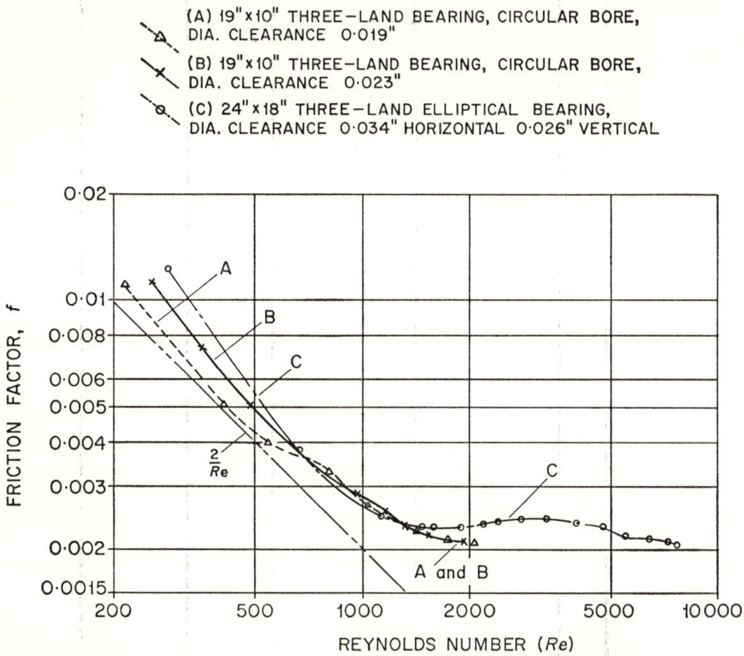

FIG. 6.10. Friction factors of three bearings tested.

It is seen that at the lower Reynolds numbers the losses are somewhat higher than the Petroff loss, but for each bearing the loss is approximately proportional to the Petroff loss over a range of Reynolds numbers. As Reynolds number increases, the ratio of test loss to Petroff loss increases, but at first only slowly; this is probably associated with nonlaminar flow in oil grooves. The loss rises more rapidly at the higher Reynolds numbers, as nonlaminar flow develops in the film, but there is no abrupt transition speed as in the hydrodynamicists' tests of wide bearings with centralized journal.

The only one of the bearings of figure 6.10 which runs far into the nonlaminar region is the 24-in. × 18-in. bearing. On this bearing, as successive nonlaminar flow regimes develop, the friction factor at first rises and then falls, exhibiting some irregularity in this variation. At the highest Reynolds numbers tested the slope of the friction-factor curve corresponds approximately to $(Re)^{-0\cdot 25}$. The test results do not justify extrapolation of the friction factor to higher Reynolds numbers.

The bearing tests plotted in this figure were carried out with constant load on each bearing. At the lower speeds the eccentricity ratios were high, and at the same Reynolds number must have been higher for bearing B with 0·023 in. clearance than for bearing A of the same design but with 0·019 in. clearance. This difference accounts for the higher value of friction factor for bearing B at the lower Reynolds numbers. The clearance ratio does not appreciably affect friction factor for Reynolds numbers above 700 in these tests.

Performance characteristics of various bearing designs

For a range of journal-bearing designs commonly used in turbo-machinery, in which clearances are normal and most of the oil supply passes through an oil film, the performance in steady running with laminar flow in the film resembles that found for the laminar regime in the 19-in. × 10-in. bearing.

The oil flow for bearings of the same overall dimensions but of different detailed design varies widely with the same oil-supply pressure. Test results showing the great effect of oil-inlet design on oil flow are given in ref. [5]. When the supply pressure is adjusted to give the same oil flow there is generally fair correspondence in loss.

For some bearings of non-standard design, in which much of the oil circulated is cooling oil which does not enter the main film, it has been found that the conventional assumptions give poorer assessment of performance than for bearings of normal design.

Reduction of bearing area in bottom half or top half generally reduces the loss, by an amount which is influenced by the grooving and oil-supply arrangements. The expedient of reducing bearing area, of course, reduces the margin of safety of the bearing under certain operating conditions.

For normal bearings which function in nonlaminar flow the available test experience shows that oil flow and loss can still be approximately forecast. The 24-in. bearings on which tests are reported in refs. [23]

and [24] run far into the nonlaminar region. These are elliptical bearings which differ in detail design of grooving and bearing surfaces. These differences have some influence on variation in loss with speed. The loss characteristics of one of these bearings is included in figure 6.10. The oil flow is approximately proportional to the speed from 1,000 r.p.m. upwards.

On these large bearings the fall in oil-outlet viscosity with increasing speed is so rapid that at constant load the duty parameter passes through a maximum at a speed in the neighbourhood of 2,000 r.p.m. As the speed is raised at constant load, the journal attitude continues to rise whether the duty parameter rises or falls. For a bearing running well into the nonlaminar region it can be seen by inspection of test results that the Reynolds number plays a more important part than the duty parameter in determining the operating conditions.

On bearings of varying clearance but otherwise of the same design the few available test results show that with constant oil-supply pressure the oil flow is approximately proportional to clearance, as would be expected. The loss characteristics of one bearing design with changed clearance are shown in figure 6.10. The practical consequence of increasing clearance with constant oil-supply pressure is that if the bearing is running in the laminar regime the oil-outlet temperature is diminished and the loss is diminished by an amount less than in inverse proportion to the clearance; while if the bearing is running in a nonlaminar regime the oil-outlet temperature is diminished and the loss is little changed.

The general conclusion is that for the bearing designs widely used in turbomachinery, so far as assessment of bearing loss is concerned, the conventional assumptions of the hydrodynamic theory, with viscosity taken as that at mean outlet-oil temperature, give reasonable approximation for bearings running in the laminar-flow regime. With nonlaminar flow, conditions are more complicated, but the Reynolds number as conventionally defined is a useful parameter in assessing performance. The oil-flow parameter for any one bearing varies little over a considerable speed range which includes both laminar and nonlaminar flow, but the value of the flow parameter is greatly influenced by details of bearing design.

CHAPTER SEVEN

Approximate Assessment of Performance

Assessment of overall performance

Approximate assessment of the overall performance of journal bearings, particularly of loss and oil flow, is required in the course of development of turbomachinery. It is convenient to have a simple and systematic means of calculating performance even when test information is available for similar bearings.

Empirical formulae have long been used for approximate assessment. The traditional formulae do not involve hydrodynamic parameters, and do not take into account all the significant influences on bearing performance.

A better assessment can be made by using the hydrodynamic approach of Chapters Four and Five. This method is well established for bearings with laminar flow in the film [4]. The full procedure involves a good deal of calculation, particularly with novel bearing designs. Only roughly approximate information is available for application of this method to bearings with nonlaminar flow in the film. For large bearings with nonlaminar flow it is preferable to assess by direct comparison with the nearest bearing already tested under similar running conditions.

Method of assessment

The method here described employs formulae of the type given by hydrodynamic theory. Bearing test and experience are examined to decide which of the numerous independent variables can be left out of account in making an assessment which is adequate for preliminary design purposes. Test observations are used to determine the values of the coefficients as functions of the variables retained.

The method assumes that the viscosity of the lubricant at the mean outlet temperature can be treated as the effective viscosity for overall performance assessment.

When oil-inlet and outlet temperatures are prescribed the effective viscosity is known and direct calculation is made of overall performance. The required oil flow is then calculated for the prescribed oil-inlet and outlet temperatures. It must be examined whether this is a reasonable oil flow for the proposed bearing design. When the oil-inlet temperature and oil flow are prescribed the loss and oil-outlet temperature have to be determined by trial calculation.

The basic formulae used in the assessment are the following from Chapter Five.

Reynolds number is defined as

$$(Re) = \frac{\rho}{\mu}U\left(\frac{C}{2}\right) = \frac{\pi}{2}\frac{\rho}{\mu}DNC. \tag{5.1}$$

Bearing loss is given by either of the alternative expressions

$$H = K_j H_p = K_j\left\{2\pi^3\mu LD^2N^2\left(\frac{D}{C}\right)\right\} \tag{5.5}$$

or

$$H = f\left(\frac{\pi^4}{2}\rho LD^4N^3\right) \tag{5.8}$$

Oil flow is

$$Q = K_q Q_R = K_q\left(\frac{\pi}{4}LDNC\right). \tag{5.6}$$

When the bearing loss is equal to the heat imparted to circulating oil the loss is related to oil-temperature rise by

$$H = J\rho c_p Qt. \tag{5.17}$$

The coefficients K_j, f, K_q are functions of the bearing design and the operating conditions. The following relation holds:

$$f = K_j\left(\frac{2}{(Re)}\right). \tag{5.7}$$

Numerical coefficients such as $2\pi^3$ appear in some of these formulae. The numerical coefficients could have been absorbed into the defined characteristic coefficients; but it is more convenient to use characteristic coefficients with the physical meanings widely adopted in analytical and experimental work. The formulae apply with any consistent units.

Many design and operating variables affect the loss and oil-flow coefficients. The assessment is simplified by restricting the application to a limited range of design and operating conditions, and by taking each characteristic coefficient as being a function only of those variables or parameters which have appreciable influence within the specified range.

The design range to which the assessment is applicable consists of bearings of the general type used in direct-coupled turbine-generators, as described in Chapter Two, but excluding the stabilized bearing design for which test information on loss and oil flow is not available. The operating range covers operation within the normal range of service loads, at speeds between half normal speed and overspeed, with lubricating oil to British Standard Specification B.S. 489 : 1968 as normally used in turbine plant, and with oil supplied at a pressure and temperature within the usual service range. The assessment will give some guidance on performance with design and operating conditions outside the specified range, but not with the same accuracy as within the specified range.

According to conventional hydrodynamic theory with unrestricted film-flow regime, the loss coefficient K_j is a function of bearing design and of the parameters S (duty parameter) and (Re) (Reynolds number). It is found by test that within the specified range of design and operating conditions the loss coefficient is little affected by change of bearing load when the other conditions remain constant. The loss coefficient can therefore be treated as independent of S. It is also found that within the specified range variation of design has a sufficiently small influence on the loss coefficient to permit of loss coefficient being treated as a function of the Reynolds number alone for approximate assessment.

The oil grade, oil-inlet temperature and oil flow substantially affect bearing loss; but in this method of assessment these variables are covered by the value of oil-outlet viscosity used in the calculation; they do not enter into the loss calculation as explicit variables.

The friction factor f, like the loss coefficient K_j, may be treated as a function of Reynolds number alone.

Figure 7.1 shows the coefficients K_j and f to be taken for approximate loss assessment. The curve is based on test results on many bearings, averaged and smoothed. Test results on only a few bearings are available at the highest Reynolds numbers. Up to $(Re) = 600$ the loss coefficient is constant with (Re); this is consistent with laminar flow in the film. The slow increase up to $(Re) = 900$ is explained by loss contribution from nonlaminar flow in grooves, and by initiation of nonlaminar flow in the film. At higher Reynolds numbers the loss coefficient is nearly proportional to (Re) and the friction factor shows relatively slow variation; there are, however, some irregularities which indicate the presence of successive nonlaminar flow regimes in the film. At the highest Reynolds numbers for which test results are available f is approximately proportional to $(Re)^{-0.25}$.

The oil-flow coefficient K_q for a given bearing with fixed oil-supply

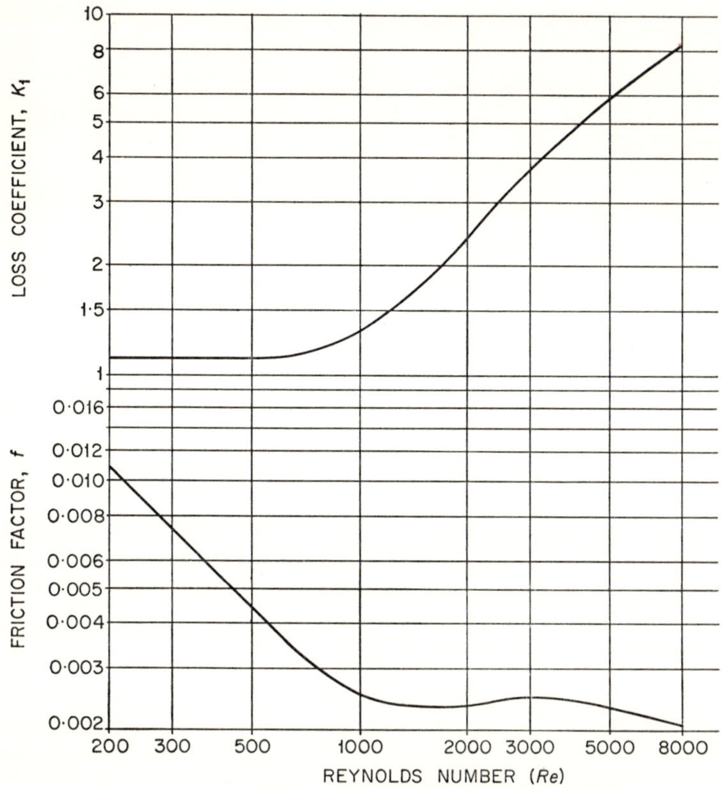

FIG. 7.1. Coefficients for approximate loss assessment.

pressure, running within the specified operating range, is only slightly affected by changes in speed, load, and oil-supply temperature. To this extent the flow coefficient may be treated for approximate assessment as independent of both S and R. The flow coefficient is, however, substantially affected by change in oil-supply pressure, and is greatly affected by change in design between one bearing and another. It is therefore not possible to quote a flow coefficient applicable to a range of bearing designs. The safest method of assessment is by comparison with test on a bearing of similar design at the same oil-supply pressure. For guidance as to order of flow coefficient it is found that for the three-land bearing design of Chapter Two, with $L/D = 0{\cdot}75$ and with oil-supply pressure 10 lb/sq in., the value of K_q is in the neighbourhood of 1·1. The flow coefficient falls with increasing L/D ratio. It is sensitive to detailed design of oil-supply grooves (especially to lead-in from groove to film), as well as to number and position of the oil-supply grooves.

In applying the results of approximate assessment to the design of lubricating-oil systems, margins must be allowed on the circulating-oil quantity and on the oil-cooler capacity. These margins must cover both the approximations in the calculation and the effects of tolerance in bearing manufacture and of variation in conditions of operation.

PART THREE

Dynamic Characteristics

CHAPTER EIGHT

Journal Bearings in Vibration

The dynamic characteristics of the journal bearings play a major part in the vibration of turbomachinery. Knowledge of these vibration effects was long purely empirical. A bearing design to avoid one type of oil whirl was described in 1925 [25]. Bearing characteristics were known to affect critical speed phenomena, but the magnitude of the change in critical speed due to the bearings was not appreciated until 1951 [26]. In recent years correlation of analytical work, dynamic experiments on bearings, and observation of machine vibration has brought about great advance in knowledge of the complex dynamic characteristics of bearings and of their relation to machine vibration.

Stability of steady-running bearing characteristics and dynamic stability of system

When the steady loading at each journal bearing is statically determinate (for example, when a rotor is carried in two journal bearings) the journal cannot run steadily unless each bearing possesses stable steady-running characteristics, which retain the journal in motion about the equilibrium position following a small disturbance. The bearing designs used in practice exhibit steady-running stability over all or the greater part of their range of operating conditions. Steady-running instability sometimes occurs with a lightly loaded bearing. The calculated characteristics of Chapter Four show steady-running instability, with attitude locus falling under reducing load, for certain designs of non-circular bearings at light loading.

Stability in steady running is a property of the bearing characteristics which is independent of the rotor characteristics.

Stability in steady running of the bearing characteristics is not sufficient to ensure stable running of a rotor in the bearings. Dynamic stability of the rotor is determined by the dynamic characteristics of the bearings and of the remainder of the system in conjunction. The rotor bearings and bearing supports must be taken into account. Stability is affected by external forces applied to the rotor.

Oil-film behaviour in vibration of journal bearings

The first step towards understanding the dynamic characteristics of a journal bearing is to know the condition of the bearing oil film in vibration. The problem of oil-film regime in vibration – the characteristic position and motion of the pressure-bearing film when the journal axis is moving relatively to the bearing axis – is separate from the problem of the hydrodynamic-flow regime within the film which has been discussed in previous chapters. Any regime of motion of the oil film may be associated with any hydrodynamic-flow regime within the film.

The experimental work in which bearing films have been observed in vibration has been carried out almost entirely on rig test, largely on small-scale rigs, and not on turbomachinery in service. The following statements as to oil-film regimes in the bearings of turbomachinery are reasonable inferences from the available evidence.

When a journal vibration of small amplitude is superimposed on the steady-running condition previously described the film regime remains generally similar to that of the steady-running condition, but change of film thickness and limited circumferential movement of the pressure-bearing film accompany the movement of the journal. If, however, the amplitude of the journal vibration becomes sufficiently large the oil-film regime changes, and it may assume various types, depending on the character of the vibration. This is one of the sources of complication of the dynamic phenomena in turbine bearings.

In examination of possible film regimes in vibration, consideration is given in the first place to an idealized bearing – a full circular bearing in which it is assumed that there are no grooves or other interruptions of the bearing surface and that sufficient oil is always available to form a film. The types of journal motion particularly examined cover any type of vibration of fairly small amplitude, and also whirl of substantial amplitude and of any frequency in the direction of rotation of the shaft. Most journal motion in turbo-rotor vibration is of one of these types.

Figure 8.1 illustrates the steady-running film regime already discussed, and shows also other regimes which may hold for the pressure-bearing film with journal in vibration. In the figure (*a*) illustrates steady film, (*b*) oscillating film, (*c*) rotating trailing film, (*d*) rotating leading film, and (*e*) rotating transition film.

A sixth regime shown in the figure as (*f*) – the continuous film – has sometimes been observed on small-scale bearing rig tests. This regime does not occur in turbine bearings as normally operated.

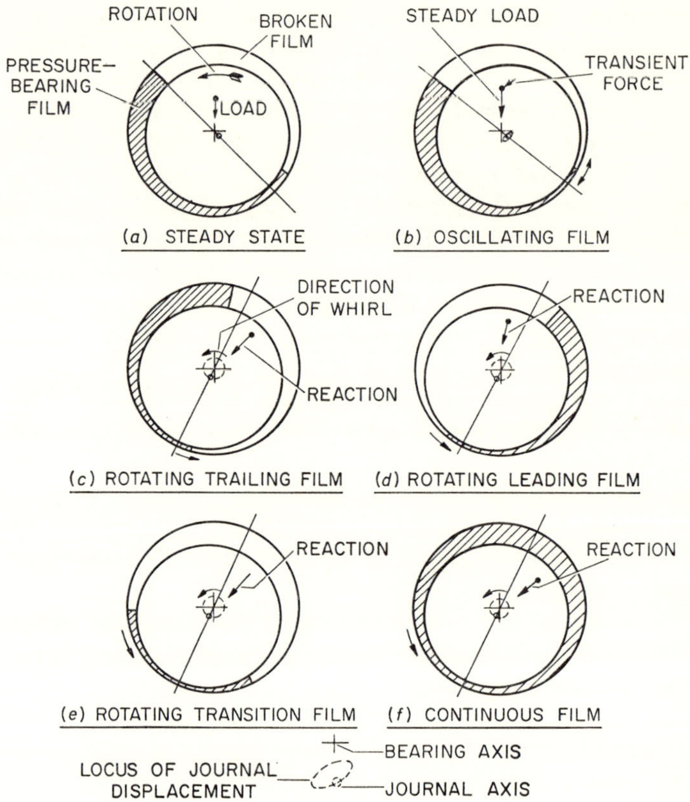

FIG. 8.1. Oil-film regimes in journal-bearing vibration.

The steady film regime, (*a*), corresponds to the steady-running condition described in earlier chapters. The film is shown as starting at the position of maximum clearance, as is often assumed in analytical investigation.

The oscillating-film condition, (*b*), is obtained when a small amplitude of vibration is superimposed on the steady-running condition; this vibration is sustained by a fluctuating force superimposed on the steady load. The motion of the journal axis results in both squeeze and angular swing of the film. This oscillating-film condition is of frequent occurrence in bearings in service. It is closely related to the steady-running condition.

In severe vibration the journal motion may become a whirl of such large amplitude that the position of minimum film thickness travels round the bearing with the whirl. In such a case the pressure-bearing film travels round with the journal. The operating condition is no longer closely related to the steady-running condition.

In the case of journal whirl of substantial magnitude at constant eccentricity the idealized bearing is in a condition of steady running relative to axes rotating with the whirl. There is an effective speed of rotation, relative to these axes, of $(N - 2F)$, where N is the speed of rotation of the journal and F is the frequency of whirl measured in the same direction. The position of formation of the pressure-bearing film depends on the direction of the mean peripheral speed of bearing and journal relative to the rotating axes, that is on whether N is greater or smaller than $2F$.

When the frequency of whirl is less than half the journal speed the pressure-bearing film is built up on the trailing side of the minimum film thickness, as in figure 8.1 (*c*). When the frequency of whirl is greater than half the journal speed the pressure-bearing film is built up on the leading side of the minimum film thickness, as in figure 8.1 (*d*). The diagrams show for each case the direction of the reaction applied by journal to bearing.

When a rotating load is superimposed on a much smaller steady load there is variation of eccentricity during each whirl rotation, but the film regimes obtained resemble those given by whirl of constant eccentricity.

With rotating trailing film, figure 8.1 (*c*), the reaction of the bearing on the journal (opposite to the reaction just mentioned) has a component normal to the attitude radius in the direction of the whirl. This component tends to increase the whirl. An explanation is thus offered of why a rotating trailing film is associated with tendency to instability. As reaction force increases and minimum film thickness decreases, the direction of the reaction swings closer to the attitude radius, so that if there is damping elsewhere in the system a steady condition of whirl may be attained.

With rotating leading film, figure 8.1 (*d*), the corresponding component tends to damp the whirl. A bearing in this regime does not excite instability.

The assumptions of the conventional hydrodynamic theory suggest that a pressure-bearing film cannot be sustained at the transition condition of whirl frequency half the speed of rotation. In practice, a pressure-bearing film sometimes forms at the transition frequency of half or approximately half the speed of rotation. A possible explanation is that on either side of the moving position of minimum film thickness the oil dragged away by relative movement of journal or bearing is warmer and of lower viscosity than the oil dragged towards this position; a pressure build-up therefore takes place on each side of the minimum film thickness. The transition condition is of importance because it may exhibit a sharply resonant tendency to maintain rotor whirl when there is a natural frequency near half running speed. Figure 8.1 (*e*) gives a diagrammatic representation of this film condition; the angular extent of the pressure-bearing film is not known.

There are additional possible transition film regimes which have been occasionally experienced in service. In some cases of transition the bearing appears to loose its grip on the journal, which whirls under comparatively low forces, although frequency is not in the neighbourhood of half the running speed.

In real bearings for turbomachinery there is always at least one hole or groove for oil admission; there is often an elaborate system of grooving with some points open to drain. There is usually a substantial bearing arc in the bottom half of the bearing, but there may be narrow or discontinuous bearing surfaces in the top half. Some bearing surface is provided in the top half as a safeguard in vibration. Turbine bearings are frequently not of circular bore. The oil-film regimes in real bearings are therefore not identical with those in the idealized circular bearing.

Nevertheless, the dynamic characteristics of a full circular bearing provide a first approximation to the dynamic characteristics of real turbine bearings, particularly in vibration under imposed disturbances. The approximation is best for vibration of small amplitude in bearings which are heavily or moderately loaded; in such cases the oil forces arise mainly in the bottom half. In whirl of large amplitude or in operation at low mean eccentricity there may be one or more films on the bearing surfaces in the top half which are discontinuous in angle and intermittent in time. In such conditions of operation the bearing design has a major influence on the dynamic characteristics. This influence must be empirically determined.

Both steady-running and dynamic characteristics of a bearing are affected when the oil flow is restricted. Diminution of oil flow leads to high oil-temperature rise in the bearing, with lower effective viscosity. In addition to this indirect effect, restriction of supply brings about a direct change of extent and conditions in the bearing-oil film, with great effect on the dynamic characteristics. Severe oil restriction may set up film cavitation, which damages the bearing without appreciable journal vibration.

Force–displacement characteristics under transient load

An introduction to some of the features of the dynamic characteristics is provided by examination of the displacement characteristics in steady

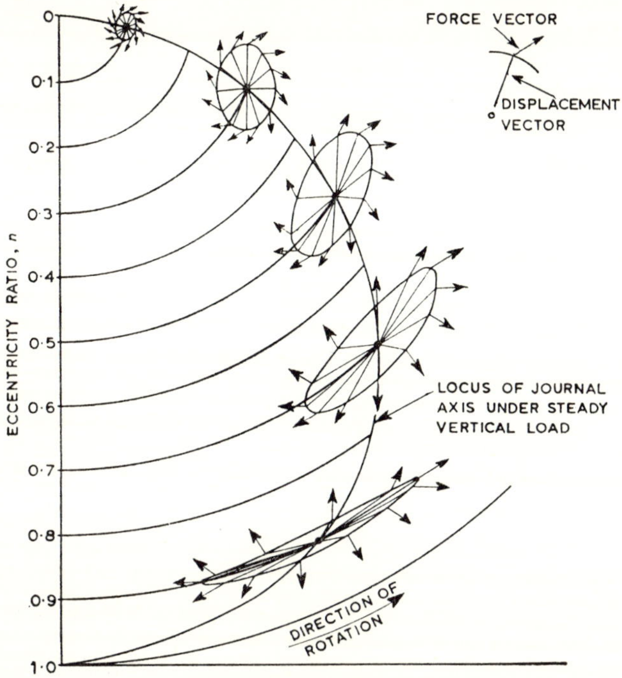

FIG. 8.2. Journal attitude diagram: displacement of journal axis in steady running.

running when transient loads are superimposed in succession on a steady load. Calculation of these characteristics as given by the conventional theory is readily carried out for the bearing of full circular bore; the characteristics of this bearing can be swung round for any angle of application of load.

Figure 8.2 shows the results of calculations of journal displacement from the steady-running attitude when transient loads of one-fifth of the steady load are superimposed on the steady load. The calculations are made for steady-load conditions giving various running attitudes, at eccentricity ratios 0·1–0·9, and for transient loads applied successively in various directions. Each transient load is assumed held long enough for steady-running conditions to be attained. The calculations are made for a full 360° bearing with $L/D = 1$, as in figure 4.3.

Inspection of the vector diagrams shows certain peculiarities in the relations between superimposed force and resulting displacement from the steady-running attitude. The characteristics are unsymmetrical, the asymmetry increasing with increasing eccentricity ratio. The force vector is inclined to the displacement vector at a variable angle. For any steady-running condition, if the force vector is slowly rotated through one complete turn, the journal axis follows a closed path; there is then input of energy to the superimposed forces with rotation in the same direction as the journal rotation, and output of energy with rotation in the opposite direction. For the lower eccentricity ratios each path of displacement is approximately an ellipse with the steady-running position as centre, showing that the characteristics are approximately linear; but with increasing eccentricity ratio, non-linearities become marked, giving a banana-shaped locus for the highest eccentricity ratio plotted.

The peculiarities associated with angle of inclination between force and displacement, and with energy input or output depending on direction of motion round a closed path, are consequences of the force–displacement relations being non-reciprocal. The tendency of the bearing characteristics to favour whirl in the direction of rotation presents a striking difference from the behaviour of elastic systems.

CHAPTER NINE

Conventional Theory of Linearized Dynamic Characteristics with Laminar Flow

Scope of analysis

This analytical investigation of dynamic characteristics starts with a reference bearing running steadily under load, in accordance with the assumptions of the conventional theory of laminar flow, as described in Chapter Four. Dynamic characteristics are introduced by imposing motion of the journal axis and examining the superimposed disturbance forces which correspond to the journal motion. Additional assumptions are made to enable the dynamic characteristics to be expressed in a linearized form which can be conveniently calculated. The accuracy of these assumptions is considered.

Dynamic characteristics thus calculated are given for certain simple bearing designs. The applicability of these results to more complicated bearing designs is considered. The calculated dynamic characteristics are applied in investigations which illustrate the effect of the characteristics on the vibration of turbomachinery.

An important application of journal-bearing dynamic characteristics, as determined by calculation or by experiment, is to the analysis of rotor motion in turbomachinery. Formal analysis of rotor motion is outside the scope of this book.

Similarity conditions in dynamic characteristics

Force–displacement relations are sufficient to determine the bearing characteristics under transient loads, as in the previous chapter, when each load is applied long enough to allow the running condition to stabilize. For rapid movements the forces on the journal are related to motion as well as to displacement. With laminar flow in the film, both displacement and velocity forces will be set up by viscous action.

One of the assumptions made for conventional calculation of steady-running characteristics is that inertia of the oil in the film is negligible. Analysis suggests that this assumption causes little error so long as laminar flow in the film is stable. If the same assumption is made for calculating dynamic characteristics it follows that the acceleration effects arising in the film are negligible. It is assumed in the present chapter that inertia of the oil is negligible; and in consequence, the dynamic characteristics of a given bearing are taken as expressed by force–displacement–velocity relations.

The conventional assumption that journal axis remains parallel to

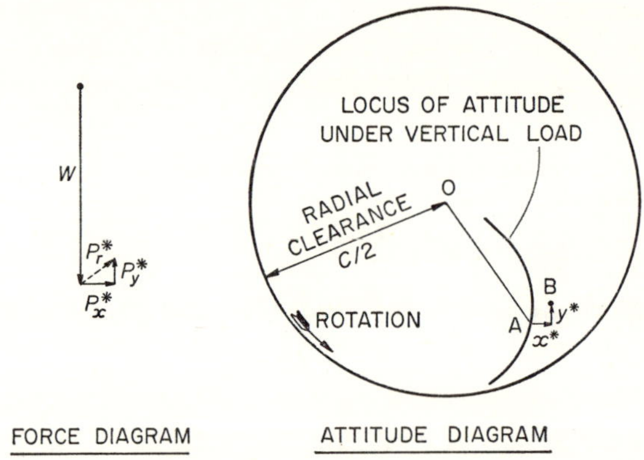

FIG. 9.1. Force and attitude in dynamic characteristics.

bearing axis is retained. Disturbance forces and journal displacements therefore take place in transverse directions, and no question arises of disturbance moments or of angular tilt of journal. 'Transverse' is used in the sense of describing any direction in the plane normal to the bearing axis.

Consider a journal running under a steady load W, on which variable disturbance forces P_x^*, P_y^* are superimposed – see figure 9.1. The disturbance forces produce displacement (x^*, y^*) of the journal axis from its steady-running position. The disturbance forces represent the loads applied through journal to bearing, measured as positive in the direction of the displacements. The disturbance force P_y^* is in the opposite direction to the steady load W.

Dimensionless expressions for disturbance forces, displacements, and velocities are given by

$$\text{Forces:} \qquad P_x = \frac{P_x^*}{W}, \quad P_y = \frac{P_y^*}{W}$$

$$\text{Displacements:} \quad x = \frac{x^*}{r}, \quad y = \frac{y^*}{r}$$

$$\text{Velocities:} \qquad \frac{\dot{x}}{\omega} = \frac{\dot{x}^*}{r\omega}, \quad \frac{\dot{y}}{\omega} = \frac{\dot{y}^*}{r\omega}$$

where r is the radial clearance and ω the angular velocity of journal, as in the analysis of Chapter Four for steady running.

If there is similarity in dynamic characteristics for bearings which exhibit Sommerfeld similarity in steady-running conditions the characteristics must take the form

$$P_x = f_1\left(x, y, \frac{\dot{x}}{\omega}, \frac{\dot{x}}{\omega}\right) \tag{9.1}$$

$$P_y = f_2\left(x, y, \frac{\dot{x}}{\omega}, \frac{\dot{y}}{\omega}\right) \tag{9.2}$$

where the functions are fixed for Sommerfeld similarity in the steady-running conditions.

The relations are clearly nonlinear when the displacement is sufficiently great to take up a substantial part of the bearing clearance.

Linearization of dynamic characteristics for small amplitude of journal vibration

In many physical phenomena the relations involving force, displacement, and velocity are linear or nearly linear for motion of small amplitude. Linear relations permit of great simplification of analysis.

It is assumed that in the bearing dynamic characteristics the force–displacement–velocity relations are linear for journal motion of small amplitude. It is not self-evident that this assumption is an accurate one. Some further consideration of this problem is given later. The ultimate check of validity of the assumption must come from correlation of the analytical results with bearing test and operating experience.

With this assumption the dynamic characteristics can be expressed in the dimensionless form

$$P_x = a_{11}x + a_{12}y + b_{11}\left(\frac{\dot{x}}{\omega}\right) + b_{12}\left(\frac{\dot{y}}{\omega}\right) \tag{9.3}$$

$$Py = a_{21}x + a_{22}y + b_{21}\left(\frac{\dot{x}}{\omega}\right) + b_{22}\left(\frac{\dot{y}}{\omega}\right) \tag{9.4}$$

where the coefficients $a_{11}, \ldots b_{22}$ are constant for conditions of Sommerfeld similarity.

If the dynamic characteristics are linear for any motion of small amplitude the limiting condition of zero velocity gives force–displacement characteristics which must be consistent with the variation in steady-running conditions. The force–displacement terms can then be calculated by differentiation of the steady-running characteristics of the bearing. This procedure, however, is valid only if the boundaries of the pressure-bearing film during vibration are determined in a particular way which is investigated later.

The conventional calculation of linear dynamic characteristics assumes that displacement from the steady-running position is small compared with the oil-film thickness and the velocity of the journal axis does not exceed that given by small-amplitude vibration with frequency of the same order as the running speed. These assumptions are additional to the assumptions for conventional calculation of the steady-running characteristics.

Quantities of various orders of magnitude have to be handled in the calculation. Compared with the diameter of the bearing, the bearing clearance and the film thickness are small quantities of the first order, while the horizontal and vertical displacements of the journal axis are small quantities of the second order. Compared with the peripheral speed of the journal, the velocity of movement of the journal axis is a small quantity of the second order. Compared with the steady load, the disturbance forces are small quantities of the first order.

Principles of calculation of linear dynamic characteristics

The differential equation determining pressure distribution in the lubricating film of a bearing in vibration, analogous to equation (4.1) for a bearing in steady running, is

$$\frac{\partial}{\partial s^*}\left(\frac{\rho h^{*3}}{\mu}\cdot\frac{\partial p^*}{\partial s^*}\right)+\frac{\partial}{\partial z^*}\left(\frac{\rho h^{*3}}{\mu}\cdot\frac{\partial p^*}{\partial z^*}\right) = 6U'\frac{\partial(\rho h^*)}{\partial s}+12\rho\frac{\partial h^*}{\partial t}. \quad (9.5)$$

Here U' is the peripheral velocity of the journal at the section considered, taking into account both journal rotation and motion of journal axis. The terms on the left-hand side are the same as in (4.1), but the right-hand side differs from (4.1) in two respects: by replacement of

U by U' and by the addition of the term in $\partial h^*/\partial t$, which represents the radial component of motion of the journal surface.

The terms on the r.h.s. represent the pumping effects due to journal rotation and vibration, while the terms on the l.h.s. are associated with the pressure gradients set up in viscous flow of the oil pumped through the film.

U' differs from $U = \pi DN$ by a small quantity of the second order.

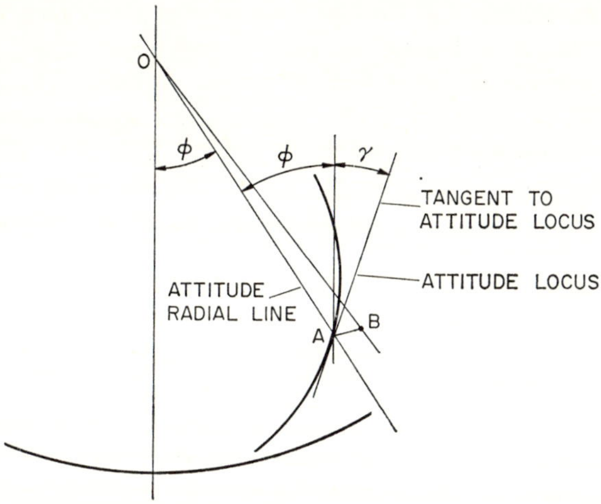

O = Bearing centre.
A = Journal centre (steady-running position).
B = Journal centre (displaced position).

FIG. 9.2. Attitude diagram.

It is permissible, for calculation of the linear dynamic characteristics, to replace U' by U in equation (9.5).

Equation (9.5) is a linear partial differential equation in p^*. The terms on the left-hand side determine variation of p^* with s^* and z^*, while the terms on the right-hand side are independent of p^*. It is therefore valid to calculate separately, and to superimpose, solutions for the terms of the r.h.s., taken one at a time, provided that the boundary conditions are fixed.

Consider the relative magnitude of the terms on r.h.s. $\partial h^*/\partial s^*$ is a small quantity of the first order and $\partial h^*/\partial t$ is a small quantity of the second order. Hence the pressures corresponding to the second term are one order smaller than the pressures corresponding to the first term.

The pressures from the first term give the steady-running force on the journal, and the dynamic displacement forces are one order smaller than the steady load. In the dynamic characteristics the disturbance forces include displacement components associated with variation in the first term and velocity components associated with the second term. These components are of the same order of magnitude when the journal motion is of frequency of the same order as the speed of rotation.

Attitude, displacement, and coordinate diagrams for the journal axis in vibration are shown in more detail in figures 9.2, 9.3, and 9.4. For dimensionless expression of the term $\partial h^*/\partial t$,

$$\frac{\partial h}{\partial t} = \frac{1}{r}\frac{\partial h^*}{\partial t} = \dot{k}\cos\beta + \dot{l}\sin\beta.$$

With the other terms rendered dimensionless as in the steady-running calculation, the following equation is obtained:

$$\frac{\partial}{\partial\beta}\left\{(1 + n\cos\beta)^3\frac{\partial p}{\partial\beta}\right\} + \left(\frac{R}{L}\right)^2\frac{\partial}{\partial z}\left\{(1 + n\cos\beta)^3\frac{\partial p}{\partial z}\right\}$$
$$= -12n\sin\beta + 24\pi\left\{\frac{\dot{k}}{\omega}\cos\beta + \frac{\dot{l}}{\omega}\sin\beta\right\} \quad (9.6)$$

This is the same as equation (4.3) except for the addition of the terms in $\dot{k}$ and $\dot{l}$, which represent the transverse velocity of the journal.

Assumptions as to the boundary condition of the pressure-bearing film are required for integration of the differential equation. The conventional assumptions are that $p = 0$ at the film boundary, with pressure zero throughout the clearance space outside the film; and that the film boundary is in the position which would be given by the conventional assumptions for the instantaneous position of the journal in steady-running conditions, with velocity terms zero. These assumptions give dynamic characteristics which are linear and which are consistent in their force–displacement terms with the terms obtained by differentiation of the steady-running characteristics of the bearing.

When the bearing arc is sufficiently short for the pressure-bearing film to start at an oil groove and to finish at an oil groove, and the pressure in both oil grooves is atmospheric, these assumed boundary conditions are closely approached in practice.

In most turbine bearings, however, the pressure-bearing film is terminated by breaking up within the bearing clearance. In this case the boundary condition in vibration is not physically defined in the way which the conventional assumptions suggest. Unless the velocity terms in the equation are zero, the value of $\partial p/\partial\beta$ is not zero at the film boundary. The effect on the characteristics is, however, small for velocity of small amplitude. The conventional assumptions are also open to the objection that in practice the film does not break up at zero (atmospheric) pressure but at a pressure which is below atmospheric and which cannot be specified by current knowledge. This lack of accurate definition of the boundary condition has more serious effects on the dynamic characteristics than on the steady-running characteristics.

The calculated dynamic characteristics are therefore of uncertain accuracy in representing the real dynamic characteristics of the bearing. The real dynamic characteristics do not necessarily closely approach linearity for small amplitude of motion of the journal.

Nevertheless, these conventional assumptions provide a convenient means of calculating dynamic characteristics which offer at least a fair approximation to the real dynamic characteristics of turbine-type bearings with laminar flow in the film.

The design of bearing for which calculation is most readily carried out is a bearing of a circular bore, with bearing arc which is a complete circle or which extends beyond the limits of the oscillating pressure-bearing film. In such a bearing the steady-running characteristics can be swung unchanged through a certain angle. Published analyses of dynamic characteristics relate almost exclusively to bearings of this type. The calculations here presented apply to circular bearings. Some general consideration is given to non-circular bearings.

For a full circular bearing the solution is a function of two parameters, the width ratio L/D (fixed by design) and the eccentricity ratio n. The following is an outline of the process of calculation.

For a bearing of specified design the film extent and the steady-running characteristics (if not already calculated) are calculated for various values of n as described in Chapter Four. The load–attitude relations are then differentiated to give the force–displacement relations for small disturbances.

Integration of the differential equation is next carried out for the terms on the r.h.s. of (9.6) in $\dot{k}$ and $\dot{l}$ – a first integration to give pressure distribution and a second integration to give resolved components of force. Detailed calculation is not required for the term in $\dot{l}$, because the independent variable β appears in the same form, $\sin\beta$, as in the steady-running term for which integration has been carried out in calculating the steady-running condition. Additional integrations are, however, required for the term in $\dot{k}$ where β appears in the form $\cos\beta$. The forces set up by journal velocity are then expressed nondimensionally as fractions of the steady load.

The linear dynamic coefficients may be referred to any chosen axes in the transverse plane. For investigation of rotor motion, horizontal and vertical axes (x, y) are usually most convenient, and the coefficients here plotted are referred to these axes. For a circular bearing, either full or partial, the calculation of the coefficients is more conveniently carried

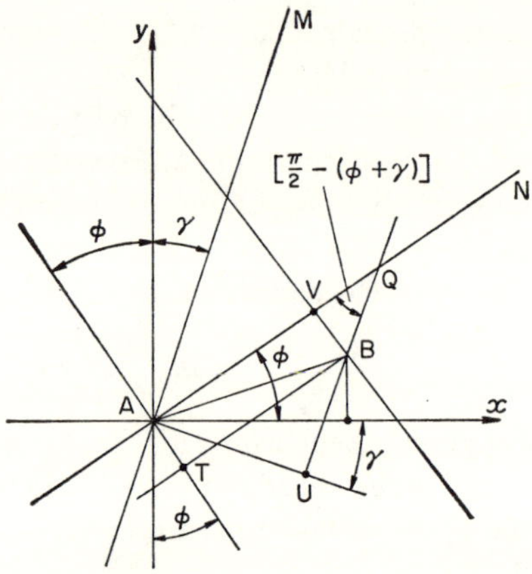

x^*, y^* = Displacements of journal centre from steady-running position.

$x = \frac{x^*}{r}, y = \frac{y^*}{r}$ = Dimensionless displacements.

x^*, y^* = Velocities of journal centre.

$u = \frac{x^*}{wr}, v = \frac{y^*}{wr}$ = Dimensionless velocities.

FIG. 9.3. Displacement diagram.

out in axes (k, l) along and normal to the attitude line. In these axes the expressions relating the coefficients to the steady-running characteristics are simpler than for other axes. The dimensionless linear relations referred to the attitude axes k, l are taken as defined by

$$P_k = a_{kk}k + a_{kl}l + b_{kk}\left(\frac{\dot{k}}{\omega}\right) + b_{kl}\left(\frac{\dot{l}}{\omega}\right)$$

$$P_l = a_{lk}k + a_{ll}l + b_{lk}\left(\frac{\dot{k}}{\omega}\right) + b_{ll}\left(\frac{\dot{l}}{\omega}\right)$$

where $a_{kk}, \ldots b_{ll}$ are constant coefficients.

The following analysis gives in more detail a procedure which establishes the dynamic characteristics in relation to the steady-running characteristics.

In figure 9.3 A is the steady running position of the journal axis and B is the displaced position (x, y). AM is the tangent to the attitude locus for vertical loading and AN is perpendicular to the attitude radius. In movement along AM, P_y changes but P_x remains zero, while in movement along AN, P_x changes but P_y remains zero. The disturbing forces corresponding to displacement (x, y) from A to B are conveniently assessed from the oblique coordinates AQ, QB, which represent the displacement of point B in (AM, AN) axes.

AU represents the projection of AB on the perpendicular to AM, and AT represents the projection of AB on the perpendicular to AN. The magnitudes of the various displacements are measured as dimensionless ratios to r.

In movement along AM, P_y changes by

$$\frac{1}{S}\left(\frac{\partial S}{\partial n}\right) \times \Delta n$$

where Δn is the change in eccentricity ratio in this movement.

In movement along AN the change in P_x is due to the load W swinging through an angle ϕ, the change in attitude angle; hence the change in P_x is equal to ϕ.

$$\text{AT} = \Delta n \text{ for step QB}$$

therefore

$$\Delta n = x \sin \phi - y \cos \phi$$

$$\text{AU} = \text{AQ} \cos (\phi + \psi) = n[\cos (\phi + \gamma)]\Delta\phi = x \cos \gamma - y \sin \gamma$$

therefore

$$\Delta\phi = \frac{1}{n}\left\{\frac{\cos \gamma}{\cos (\phi + \gamma)}x - \frac{\sin \gamma}{\cos (\phi + \gamma)}y\right\}.$$

Hence the values of the coefficients are

$$\left.\begin{aligned} a_{11} &= \frac{1}{n}\frac{\cos \gamma}{\cos (\phi + \gamma)}, & a_{12} &= \frac{1}{n}\frac{\sin \gamma}{\cos (\phi + \gamma)} \\ a_{21} &= \frac{1}{S}\frac{\partial S}{\partial n} \sin \phi, & a_{22} &= -\frac{1}{S}\frac{\partial S}{\partial n}\cos \phi \end{aligned}\right\} \qquad (9.7)$$

In this way the displacement coefficients referred to (x, y) axes have been established by a method which illustrates their relation to the steady-running characteristics.

The displacement coefficients referred to attitude axes, as given by differentiation, are

$$\left.\begin{aligned} a_{kk} &= S\frac{\partial}{\partial n}\left(\frac{\cos\phi}{S}\right), & a_{kl} &= \frac{\sin\phi}{n} \\ a_{lk} &= -S\frac{\partial}{\partial n}\left(\frac{\sin\phi}{S}\right), & a_{ll} &= \frac{\cos\phi}{n} \end{aligned}\right\} \tag{9.8}$$

These two sets of coefficients – those referred to the (x, y) axes and those referred to the (k, l) axes – are consistent with the transformation relations given in Appendix III.

For the terms involving the journal velocities the dimensionless relations expressed in the attitude axes are given in the form

$$P_k = b_{kk}\left(\frac{\dot{k}}{\omega}\right) + b_{kl}\left(\frac{\dot{l}}{\omega}\right)$$

$$P_l = b_{lk}\left(\frac{\dot{k}}{\omega}\right) + b_{ll}\left(\frac{\dot{l}}{\omega}\right).$$

The process of integration gives $b_{kl} = b_{lk}$. This is consistent with the general theory of viscous damping. The following relations for the circular bearing are established in ref. [27].

$$b_{kl} = b_{lk} = -\frac{2\cos\phi}{n} = -2a_{ll} \tag{9.9}$$

$$b_{ll} = \frac{2\sin\phi}{n} = 2a_{kl}. \tag{9.10}$$

For a circular bearing in which the bearing surface extends beyond the pressure-bearing film the calculation of all the coefficients referred to attitude axes, except b_{kk}, is therefore relatively simple if the steady-running characteristics have already been calculated. A separate integration, however, is required to determine the coefficient b_{kk}.

It has been shown (Appendix II of ref. [27]) that for a narrow bearing in which the Ocvirk condition that $\partial p/\partial z$ is of smaller order than $\partial p/\partial s$ applies, the following relation holds:

$$b_{kk} = -2a_{lk}. \tag{9.11}$$

Analytical expressions for the coefficients calculated for the very narrow bearing using the Ocvirk theory are given in Appendix I.

For the case of the very wide bearing, the analytical expressions for the coefficients are complicated. The coefficients are best calculated numerically from the steady-running characteristics. The formula for the coefficient b_{kk}, which cannot be calculated from the steady-running characteristics, is given in Appendix II.

Dynamic characteristics calculated by conventional theory

The calculated coefficients for the dynamic characteristics of full circular bearings are reproduced from ref. [28]. The calculations cover various length/diameter (L/D) ratios. The calculations for very narrow bearings are based on the Ocvirk theory [11] and for very wide bearings

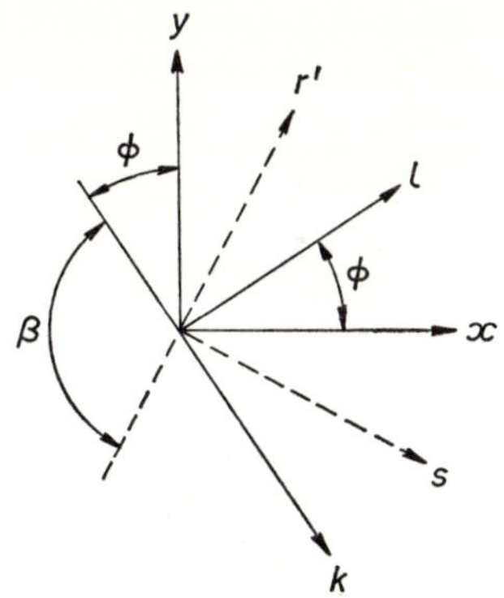

Dimensionless Coordinates.

x, y = Horizontal and vertical.
k, l = Along and normal to attitude line.
s, r' = Tangential and radial at local section of oil film.
z = Axial (perpendicular to xy plane).

FIG. 9.4. Coordinate axes.

on the steady-running characteristics calculated by Cameron and Wood [10]. For bearings of intermediate width the steady-running characteristics of refs. [8] and [10] are used, supplemented by a method of interpolation to complete the calculation of the velocity coefficients as explained in ref. [28].

At low eccentricity ratios there is slow convergence and divergence of the oil film, and the boundary conditions conventionally assumed for the pressure-bearing film are not realistic. Coefficients are not plotted for eccentricity ratios below 0·1; they must be regarded as of doubtful accuracy for eccentricity ratios slightly exceeding 0·1.

At eccentricity ratios approaching unity the characteristics for very wide bearings and for very narrow bearings tend to limiting values

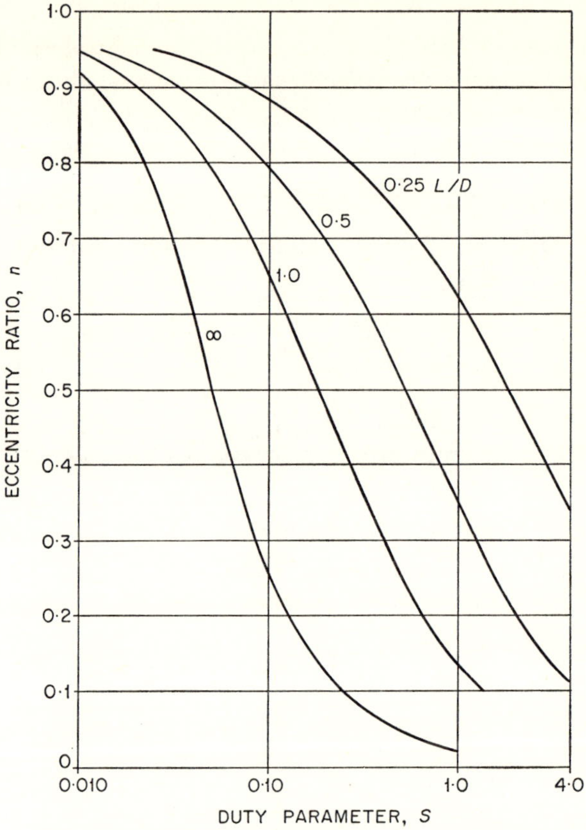

FIG. 9.5. Eccenticity ratio and duty parameter.

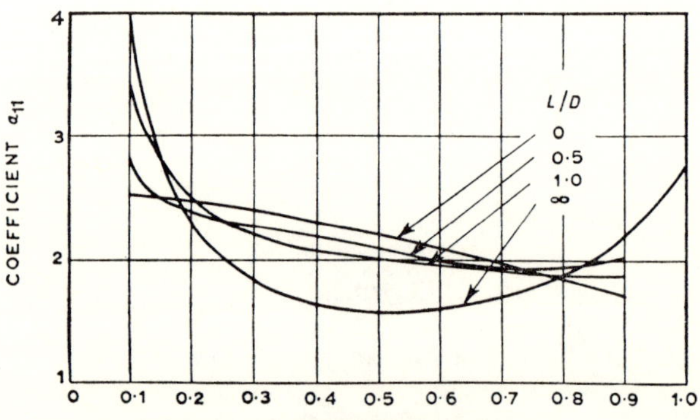

FIG. 9.6. Coefficient a_{11}.

which are given in Appendix IV. In a bearing of normal width at very high eccentricity ratio the circumferential pressure gradients in the oil film are of more significance than the axial pressure gradients in determining the oil-pressure distribution, and the relations for a very wide

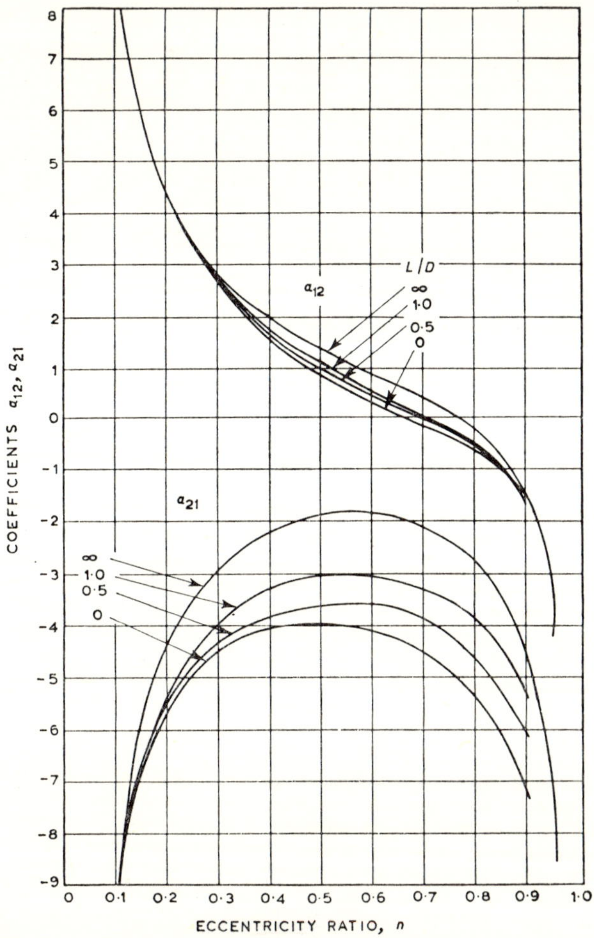

FIG. 9.7. Coefficients a_{12}, a_{21}.

bearing therefore provide a first approximation to those for a bearing of finite width.

The above procedure has permitted the calculation, without excessive labour, of a full set of coefficients which are of sufficient accuracy for

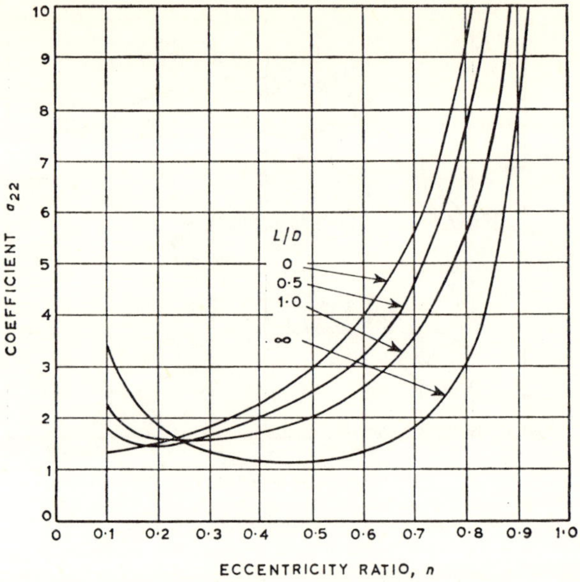

FIG. 9.8. Coefficient a_{22}.

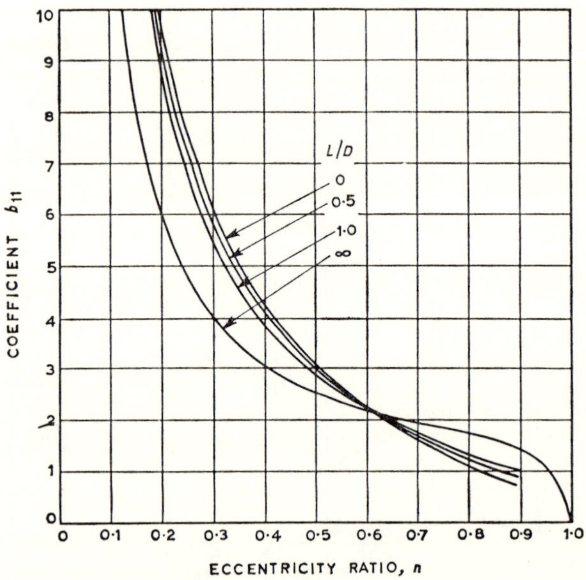

FIG. 9.9. Coefficient b_{11}.

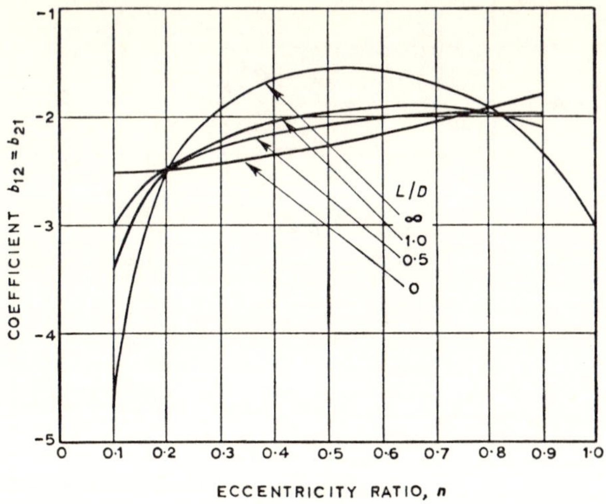

FIG. 9.10. Coefficient $b_{12} = b_{21}$.

application in many engineering problems involving bearings in the laminar-flow regime.

The calculation of the coefficients for varying values of L/D shows that each coefficient exhibits wide variation at constant duty parameter, but only small variation at constant eccentricity. The coefficients are therefore plotted as functions of eccentricity. The relation between duty

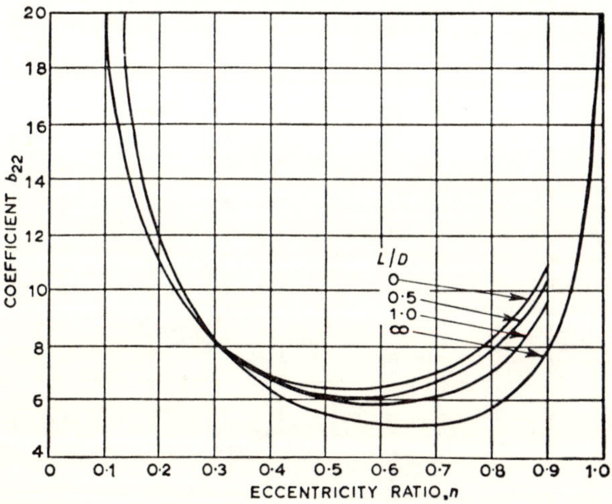

FIG. 9.11. Coefficient b_{22}.

parameter and eccentricity, given by the conventional theory of steady running, is shown in figure 9.5. The dynamic coefficients are plotted in figures 9.6–9.11. All these calculations apply to full circular bearings.

Applicability of calculations to some turbine bearing designs

In real turbine bearings of circular bore the bearing surface is interrupted by grooving. Calculations of steady-running conditions show that if there is a substantial bearing arc in the bottom half the characteristics at high and moderate eccentricity ratios are nearly the same as for a full bearing. It is reasonable to assume that this also applies to the dynamic characteristics for motion of small amplitude.

A non-circular design of bearing often used in turbines is the 'elliptical' type, with circular bearing arc in top and bottom but with vertical clearance less than horizontal clearance. Calculations of steady-running conditions for bearings of this design have shown that for heavy loading conditions giving small minimum thickness in the bottom half the attitude relative to bottom bearing arc is the same or nearly the same as for a circular bearing. With moderate and light loading, however, substantial pressure-bearing films are formed in both top and bottom halves. In consequence, for a lightly loaded 'elliptical' bearing both steady-running and dynamic characteristics differ significantly from those calculated for a circular bearing.

For a bearing in which the surface is interrupted by grooves, or is not of circular bore, it is possible to calculate the dynamic characteristics by application of the conventional assumptions. The procedure is much more elaborate than for a full circular bearing, which has characteristics which can be swung about the bearing axis. In the non-circular bearing load–attitude relations have to be calculated for variations both in attitude radius and in attitude angle. The attitude locus under vertical load is then determined and the field of steady-running characteristics is differentiated to obtain the displacement coefficients. Separate integration of pressure variations in the oil film set up by applied journal velocity is required to obtain velocity coefficients.

Calculations of this nature are rarely carried out. Such an elaborate method is of doubtful value in view of the various uncertainties associated with the conventional calculation of dynamic characteristics. The dynamic characteristics of non-circular bearings are usually assessed by comparison with calculated characteristics of circular bearings, or from rig test results when available.

Conditions for stability in steady running and positive damping

As mentioned in the previous chapter, steady running of a rotor carried in two journal bearings cannot be obtained unless both bearings possess stable steady-running characteristics under the applied load. Without this stability the journal is not located in its equilibrium position. Stability is obtained when the bearing characteristics give a restoring component of force for every transverse displacement of the journal from its equilibrium running position. Investigation by transformation of axes (Appendix III) shows that restoring components in every orientation are given when

$$a_{11}, a_{22}, \text{ and } \left\{a_{11}a_{22} - \left(\frac{a_{12} + a_{21}}{2}\right)^2\right\}$$

are all positive.

This condition is not sufficient to ensure dynamic stability of the rotor. The conditions for dynamic stability do not depend on the bearing characteristics alone, and must be established by analysis of rotor motion.

Stability of characteristics in steady running is indicated by calculation, using the conventional assumptions, for the whole range of operating conditions in full circular bearings. It must be remembered, however, that at low eccentricity ratios which correspond to light loading there is only small divergence and convergence of the oil clearance, and the conventional assumed boundary conditions of the film are not realistic. There may be static instability in real circular bearings at low loading. In 'elliptical' bearings of the usual ellipticity ratio conventional calculation shows instability in steady running at light loading. Static stability is calculated at moderate and high eccentricity ratios for all normal designs of turbine bearings.

Physical consideration shows that with force–velocity characteristics set up by viscous action in the film, damping must be positive and reciprocal for every direction of transverse displacement of the journal. Hence in the characteristics calculated with laminar flow in the film for bearing designs in which the conventional assumptions hold,

$$b_{12} = b_{21};$$

and

$$b_{11}, b_{22}, \text{ and } (b_{11}b_{22} - b_{12}^2)$$

are positive.

Response to disturbance of frequency equal to speed

The imposed disturbance forces at turbine journal bearings are usually predominantly of frequency equal to speed. Out-of-balance is the commonest source of such disturbance in rotors, but, in general, out-of-balance in a rotor does not result in pure out-of-balance forces at the

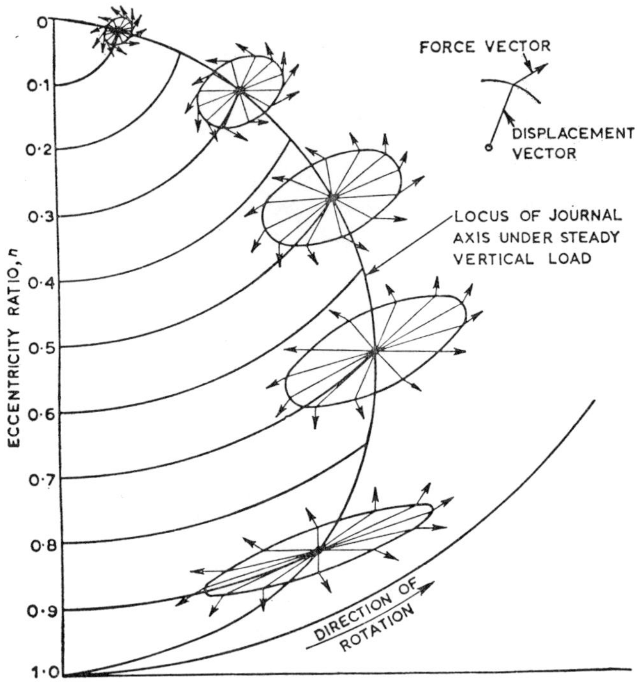

FIG. 9.12. Journal attitude diagram: response to out-of-balance disturbance.

journals. Pure out-of-balance force exhibits circular symmetry; but lack of symmetry in the bearing characteristics and in the bearing supports causes the rotor to move in an elliptical orbit, and consequently the rotor acceleration forces depart from circular symmetry.

The linearized characteristics conventionally calculated for full circular bearings are applied in calculating response to two simple types of disturbance – constant disturbance force rotating with the journal (pure out-of-balance force) and constant disturbance force rotating at the same speed in the opposite direction. The latter disturbance is unlikely to occur in practice, but is chosen as a limiting case. In consequence of the non-reciprocal nature of the bearing dynamic characteristics, these two cases give quite different journal response. Provided the relations are linear, all systems of transverse disturbance of synchronous frequency can be represented by superposition of these two limiting systems in appropriate amplitude and phase.

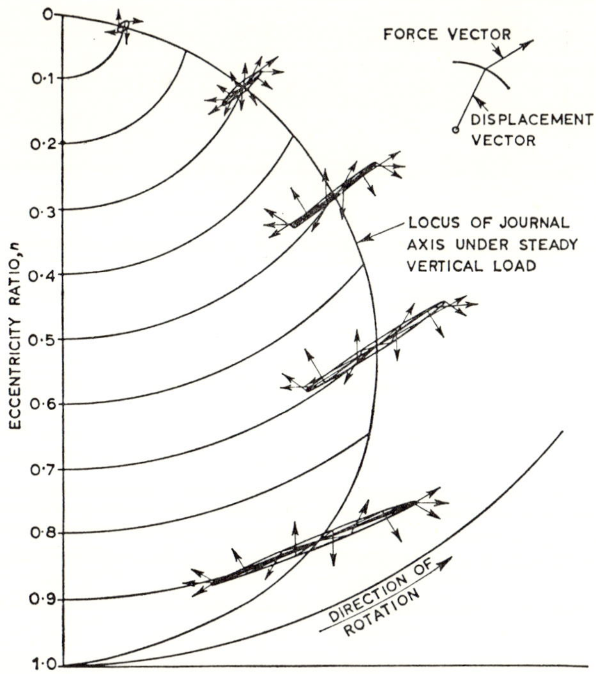

FIG. 9.13. Journal attitude diagram: response to disturbance in reverse rotation.

For each case plotted the rotating disturbance force is one-fifth of the steady load. Figure 9.12 shows the calculated paths for disturbance force rotating with the journal as in out-of-balance. The paths are given for five different steady-running conditions of varying eccentricity ratio. The vectors of force and displacement illustrate how the force generally leads the displacement in phase.

The corresponding paths are plotted in figure 9.13 for the case of constant force in reverse rotation at the same speed. It is seen that the journal motion about each steady-running attitude is nearly in a straight line.

Comparison of the dynamic out-of-balance motion of figure 9.12 with the displacement in steady running of figure 8.2 shows that the phase relations between force and displacement are altogether different for these two cases, but there is little change in the maximum amplitude of displacement under the same disturbing force.

Brief consideration is given to the effect of these bearing characteristics on rotor motion under out-of-balance. Even for this simple condition of disturbance, the rotor motion is complicated. The journal displacement is out of phase with the disturbance force. When the rotor shaft is flexible the bend of the shaft is not coplanar with the journal displacement. Journal-bearing dynamic characteristics are unsymmetrical. There is often also lack of symmetry in the bearing supports, with principal axes in a different direction. The inertia forces arising in the rotor are determined by the resultant elliptical motion of the rotor. Coupling effects between transverse directions may be excited by gyroscopic action in the rotor as well as by the bearing characteristics.

The motion of the rotor is affected if the bearings are run at altered duty parameter. This can be brought about by change of oil viscosity in a bearing running at constant speed and constant load; but the calculations show that a limited change of viscosity has only a small effect on the dynamic characteristics.

Even for the simple case of a single-span rotor carried in two similar bearings with equal load distribution, each critical speed, which with simple support at the bearings would have exhibited a single peak, is replaced by two critical speeds when allowance is made for bearing dynamic characteristics. Both these critical speeds are lower than the critical speed calculated for simple support; each critical speed is subject to appreciable damping. The two critical speeds may be submerged in a single speed range of vibration of increased amplitude.

Influence of journal bearings on rotor stability

The stability of the system consisting of a rotor running in bearings of stable force–displacement characteristics is determined by the dynamic characteristics of the whole system. In unstable running there is a net input of energy to the vibrating system which tends to increase the amplitude of vibration. When the instability is due to bearing oil whirl input of energy takes place in the journal bearings. An examination of the circumstances in which input of energy to the oscillating journal takes place therefore assists in understanding the conditions which favour rotor instability.

It is rarely the case that all the bearings in a turbine are similar and similarly loaded, and stability will depend on the balance between energy input and energy output over a number of bearings and other components. A generator rotor, however, is usually carried in two similar bearings which may show transition at the same running condition.

The case of whirl of high amplitude in a circular bearing has already been mentioned. It was pointed out that for motion against the direction of rotation, or for motion with frequency over half the running speed in the direction of rotation, a rotating leading film is formed which provides damping and thus promotes stability. For motion in direction of rotation with frequency below half the running speed, a rotating trailing film is formed which tends to put energy into the whirl.

In journal vibration of small amplitude there is no physical discontinuity of oil-film conditions associated with a particular frequency of vibration. There may, however, be energy input or energy output at the journal, depending on the orbit and the frequency of vibration as well as on the bearing characteristics. The dynamic characteristics calculated for the full circular bearing are used to establish the conditions which determine whether there is energy input or energy output. The calculation is made for steady journal motion of one frequency component; the orbit is then elliptical or circular.

Harmonic motion of the journal with frequency f is represented in its most general form by

$$x = X \cos (ft - \delta_1)$$
$$y = Y \cos (ft - \delta_2)$$

where X, Y are constants.

The net work done by the applied forces when the journal axis moves once round its elliptical path is

$$E = \int P_x \mathrm{d}x + \int P_y \mathrm{d}y \tag{9.12}$$

with the integrals taken once round the path.

Integration gives

$$E = -A(a_{12} - a_{21}) + \pi\frac{f}{\omega}\{X^2 b_{11} + 2XYb_{12} \cos (\delta_2 - \delta_1) + Y^2 b_{22}\} \tag{9.13}$$

where A is the area enclosed by the path.

The expression for E includes two terms:

(1) a 'whirl' term, $-A(a_{12}-a_{21})$, independent of the remaining coefficients and of frequency; this term is negative for A positive, that is for whirl in the direction of rotation, and positive for whirl in the opposite direction;

(2) a 'damping' term, proportional to f/ω and involving only the velocity coefficients; this term is always positive.

Hence for a given orbit of motion of the journal input of energy to the vibration (negative value of E) is obtained with movement of sufficiently low frequency in the direction of rotation. At the transition frequency ratio of f/ω between energy output and energy input, $E=0$.

The transition frequency ratio for the general case is

$$\frac{f}{\omega}=\frac{(a_{12}-a_{21})\sin(\delta_2-\delta_1)}{(X/Y)b_{11}+2b_{12}\cos(\delta_2-\delta_1)+(Y/X)b_{22}}. \qquad (9.14)$$

With constant dynamic coefficients and varying orbit of motion, the highest value of transition frequency ratio is

$$\frac{f}{\omega}=\frac{a_{12}-a_{21}}{2\sqrt{(b_{11}b_{22}-b_{12}^2)}}. \qquad (9.15)$$

This highest value is obtained when

$$\left(\frac{X}{Y}\right)=\sqrt{\left(\frac{b_{22}}{b_{11}}\right)} \text{ and } \cos(\delta_2-\delta_1)=-\frac{b_{12}}{\sqrt{(b_{11}b_{22})}}. \qquad (9.16)$$

When the motion is a circular whirl, $X=Y$ and $(\delta_2-\delta_1)=\frac{1}{2}\pi$. The transition frequency ratio for circular whirl is then

$$\frac{f}{\omega}=\frac{a_{12}-a_{21}}{b_{11}+b_{22}}. \qquad (9.17)$$

For narrow bearings of full circular bore, this has the value $\frac{1}{2}$ at all eccentricities; but for circular bearings of finite width the value is less than $\frac{1}{2}$.

These relations show that for harmonic motion in a given orbit in the direction of rotation there is a transition frequency above which there is energy output from the whirl and below which there is energy input to the whirl; that is, high-frequency motion favours stability and low-frequency motion favours instability. Motion against rotation always favours stability.

Figure 9.14 shows the transition frequency ratio in circular whirl. Figure 9.15 shows the limiting transition frequency ratio in elliptical

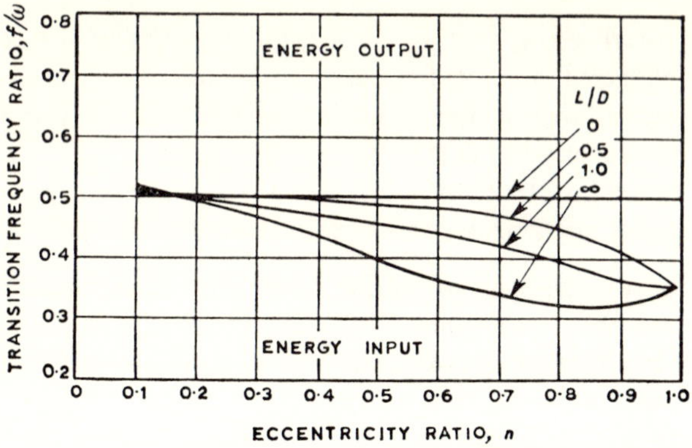

FIG. 9.14. Transition frequency ratio in circular whirl.

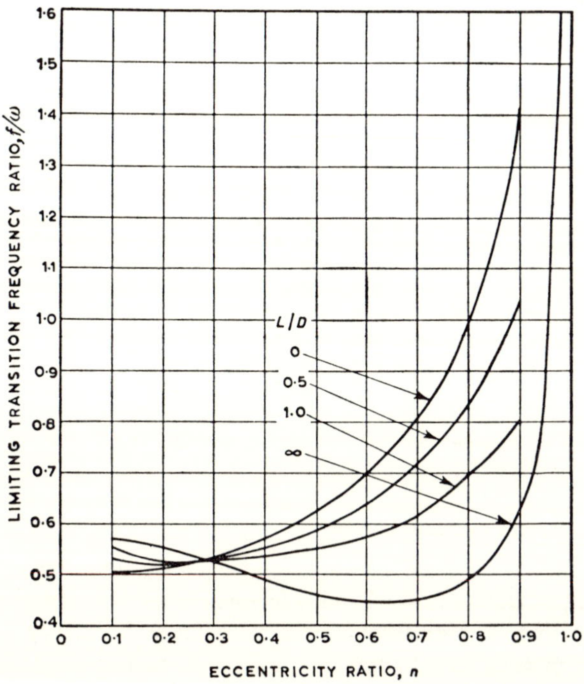

FIG. 9.15. Limiting transition frequency ratio in elliptical whirl.

whirl, in the orbit which gives the highest transition frequency ratio for each steady-running condition. Figure 9.16 shows the orbit of motion for highest transition frequency ratio; in this figure the relative size of the ellipses is of no significance.

It is seen that at moderate eccentricities, such as generally hold in

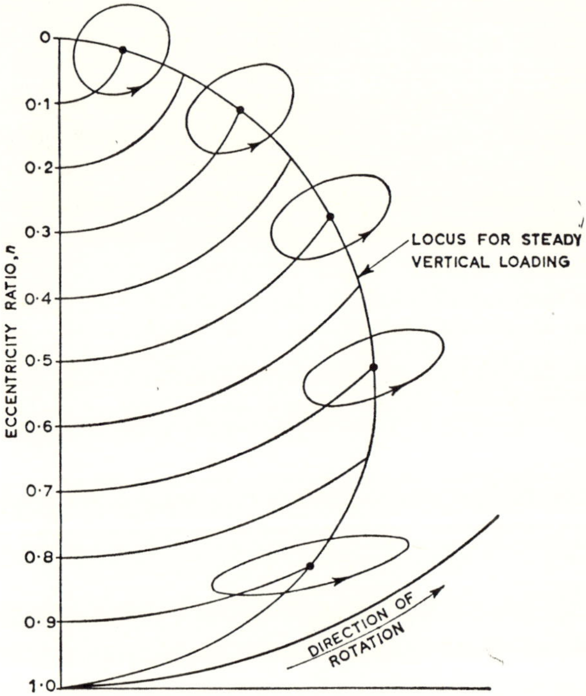

FIG. 9.16. Journal attitude diagram: orbit of motion at limiting transition frequency ratio.

normal turbine bearing operation, the transition frequency ratio for circular whirl is below 0·5 for bearings of normal width, but ratios well above 0·5 are obtained with certain orbits of elliptical motion.

For journal vibration in a given orbit, but with varying frequency above the transition frequency, the energy absorption at the bearing falls with falling frequency. If therefore a rotor is exposed to excitation elsewhere than at the bearings instability is readily excited at frequencies slightly exceeding the transition frequency for the bearings alone.

Figures 9.14, 9.15, and 9.16 apply only to journal motion of small amplitude, where the bearing characteristics can be taken as linear. At intermediate amplitudes the bearing characteristics are nonlinear, and analytical investigation becomes extremely complicated. There is increasing restraint of motion as the bearing clearances are taken up. There is also discontinuity in the oil-film conditions at a certain amplitude of whirl. Bearing oil whirl which has grown up from low amplitude is held at some amplitude – intermediate or large – by non-linearity in the bearings or elsewhere in the system.

The force–displacement coefficients a_{11}, a_{22} do not appear explicitly in the expression (9.13) for energy absorption at the bearings. In practice, these coefficients greatly influence dynamic stability of rotors, through their influence on frequency of unstable vibration and on configuration of the journal orbit.

This energy method of analysis is significant only in those stability problems where the transition condition is represented by harmonic motion of the journal in an orbit. Nevertheless, the investigation illustrates the influence of the bearing dynamic characteristics in many rotor stability problems.

CHAPTER TEN

Effects of Inertia of Oil on Dynamic Characteristics

In calculation of the steady-running characteristics of bearings it has been found that so long as the film-flow regime is laminar, little error is introduced by assuming that the inertia of the oil is negligible. The inertia of the oil must, however, be taken into account in assessing the liability to instability of the laminar-flow regime, and in making calculations for nonlaminar flow conditions.

The assumption of negligible inertia of oil, as made in the conventional calculation of characteristics in steady running with laminar flow, was retained in the previous chapter for the conventional calculation of dynamic characteristics with laminar flow in the film.

The observation that inertia effects on steady-running characteristics are small with laminar flow does not imply that inertia effects on dynamic characteristics are necessarily small with laminar flow in the film. The validity of this assumption requires separate examination. An investigation made in ref. [29] is here summarized.

Hydrodynamic consideration of the forces on an oscillating body in a moving incompressible fluid shows that in general the displacement, velocity, and acceleration of the body affect the fluid reactions on the body. Applying these conclusions to the journal bearing, and retaining the assumption that dynamic characteristics are linear in journal vibration of small amplitude, the dynamic characteristics for transverse motion of small amplitude take the form

$$P_x = a_{11}x + a_{12}y + b_{11}\left(\frac{\dot{x}}{\omega}\right) + b_{12}\left(\frac{\dot{y}}{\omega}\right) + c_{11}\left(\frac{\ddot{x}}{\omega^2}\right) + c_{12}\left(\frac{\ddot{y}}{\omega^2}\right) \quad (10.1)$$

$$P_y = a_{21}x + a_{22}y + b_{21}\left(\frac{\dot{x}}{\omega}\right) + b_{22}\left(\frac{\dot{y}}{\omega}\right) + c_{21}\left(\frac{\ddot{x}}{\omega^2}\right) = c_{22}\left(\frac{\ddot{y}}{\omega^2}\right) \quad (10.2)$$

These relations differ from (9.3) and (9.4) by the inclusion of acceleration terms. The variables $(\ddot{x}/\omega^2)$, $(\ddot{y}/\omega^2)$ provide dimensionless measure of acceleration. The assumption of linearity in the relations glosses over various complications, such as the possibility of coupling between displacement, velocity, and acceleration terms, but greatly facilitates approximate calculation of the inertia effects.

These relations (10.1) and (10.2) are not restricted to film flow in the laminar regime. Similarity of the non-dimensional dynamic characteristics, expressed by all the coefficients $a_{11}, \ldots c_{22}$ being constant, is obtained only when the Reynolds conditions for hydrodynamic similarity hold good – that is, when there is geometric similarity of the flow passages (including the same clearance ratio and the same journal attitude) and the bearings compared are running at the same Reynolds number. The conditions for Reynolds similarity are more restricted than those for Sommerfeld similarity of bearings in viscous flow, as was discussed in Chapter Five.

Inspection of the equations shows that for bearings with laminar flow in the film running at the same Sommerfeld similarity conditions the acceleration coefficients $c_{11}, \ldots c_{22}$ cannot be independent of Reynolds number, although the other coefficients are independent of Reynolds number. This is seen by considering the operation of two bearings of identical design and loading with lubricants of the same viscosity but of different density. In such a case the acceleration coefficients must be proportional to the density of lubricant, and therefore to the Reynolds number.

The acceleration terms have been calculated for bearings of circular clearance, with laminar flow in the film, as was done for the displacement and velocity terms in the previous chapter.

The linearized relations permit of separate evaluation of each group of terms; the coefficients of one variable can be calculated assuming that all the other independent variables are zero. Coefficients previously calculated for the displacement and velocity terms therefore continue to hold good when the acceleration terms are taken into account, and the coefficients of each acceleration term can be calculated by assuming a journal acceleration imposed on the steady-running condition. The calculation is made with assumptions similar to those used in the calculation of displacement and velocity terms. Results of the same order of accuracy may be expected for operation with laminar flow.

When acceleration is imposed on a journal in steady running the acceleration produces a 'squeeze' of the pressure-bearing film. The oil in the film is accelerated circumferentially and axially in making way for the journal. The oil acceleration is effected by pressure gradients in

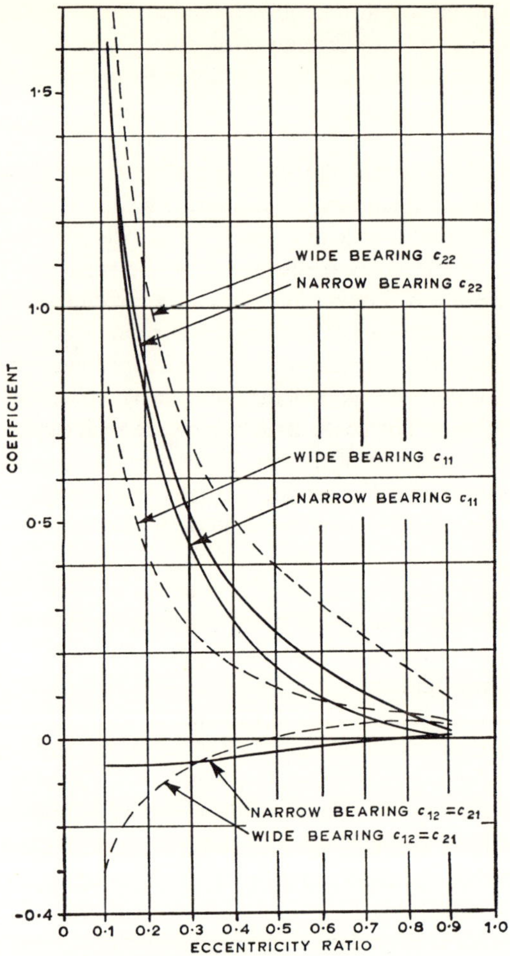

Calculated for $(Re)(C/D) = 1$.

FIG. 10.1. Force–acceleration coefficients for wide and narrow bearings.

the film. There is also radial acceleration of the oil, but the radial acceleration is of smaller order and is maintained over a radial distance of smaller order. The radial acceleration, and the radial variation of the other accelerations, are neglected in the calculation.

Negative pressures in the film are accepted in the acceleration calculation because these pressures are smaller than the pressures in steady running on which they are superimposed. The film boundaries are taken to be at zero reference pressure, that is, at atmospheric pressure, and to be unchanged in position by application of acceleration (as was done in calculation of the velocity terms).

Solution of a partial differential equation is required to establish the pressure distribution and the corresponding acceleration distribution with the assumed boundary conditions. The force on the journal is found from the resultant load of the pressure distribution.

The steps of the analysis are given in the reference quoted. It is found that for given Sommerfeld similarity conditions the coefficients $c_{11}, \ldots c_{22}$ are not constant but are proportional to $(Re)(C/D)$, where (Re) is the Reynolds number as defined in Chapter Five. This is in agreement with Cole's parameter for inertia effects in the steady-running condition. The coefficients c_{12}, c_{21} are reciprocal.

Numerical calculation of the acceleration coefficients has been made for wide bearings and for narrow bearings. The results are plotted in figure 10.1. It is found that at the same value of eccentricity ratio and of $(Re)(C/D)$, the values of the acceleration coefficients for wide and for narrow bearings are fairly near to one another. It is reasonable to expect that the values for bearings of finite width fall within the same band. The numerical values have been plotted for $(Re)(C/D) = 1$, which is a little below the highest value likely to be attained in a bearing of turbine type with laminar flow in the film.

The numerical values of the acceleration coefficients are much less than those of the displacement and velocity coefficients. Hence with vibration of frequency in the neighbourhood of the running speed the acceleration force components are smaller than the displacement and velocity force components, so long as the bearing is operating under laminar-flow conditions.

The lubricant acceleration terms may be looked at from another point of view, as adding virtual inertia to the journal. This viewpoint is convenient for examining the effect of the oil acceleration in certain rotor vibration problems. The virtual inertia added is different in different transverse directions. The angle of inclination of principal axes of virtual inertia is shown in figure 10.2.

It is found that for bearings of the material and proportions normally used in turbines the virtual inertia of the oil film is a multiple of the inertia of the journal. This virtual inertia ratio is high for a wide bearing (high L/D), and approaches zero for a narrow bearing. The virtual inertia ratio is inversely proportional to clearance ratio C/D. With journal running at the same attitude, it is independent of viscosity. Calculated values of the virtual inertia ratio in running at constant eccentricity ratio, for an oil-lubricated bearing with steel journal, with clearance ratio 0·001, are shown in figure 10.3.

Whether this virtual inertia has a substantial effect on the rotor

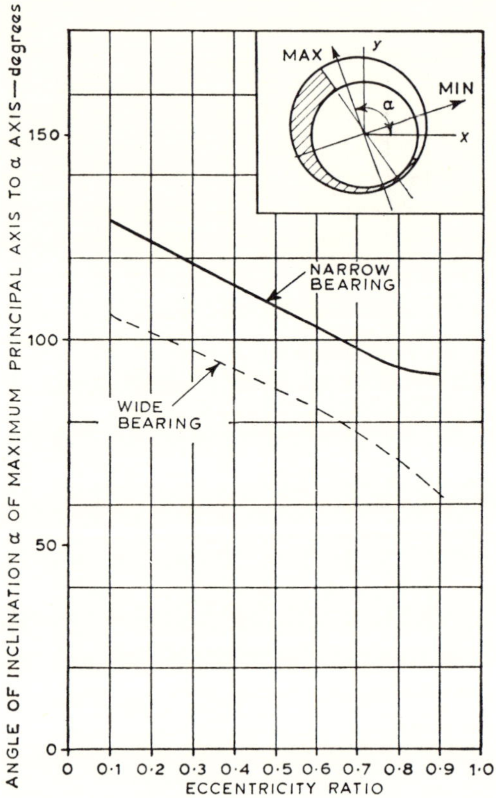

FIG. 10.2. Angle of inclination of principal axes of virtual inertia.

motion depends on the system and the mode of vibration. In most turbine rotors the mass of the rotor is a high multiple of the mass of the journals, and in rotor vibration the average amplitude of this mass generally exceeds the amplitude of the journals. In such a case the virtual inertia of the oil film is likely to have little effect on the motion as a whole. There are, however, cases of short, rigid rotors where the total mass does not greatly exceed the mass of the journals. The rotors used in bearing test rigs are sometimes of this type. In such a case the inertia of the oil plays a greater part in the motion than the inertia of the rotor itself, and it may have an important influence on the test results.

When the bearing is running with nonlaminar flow in the film the equations (10.1) and (10.2) still represent the linearized characteristics for conditions of Reynolds similarity. There is at present no established

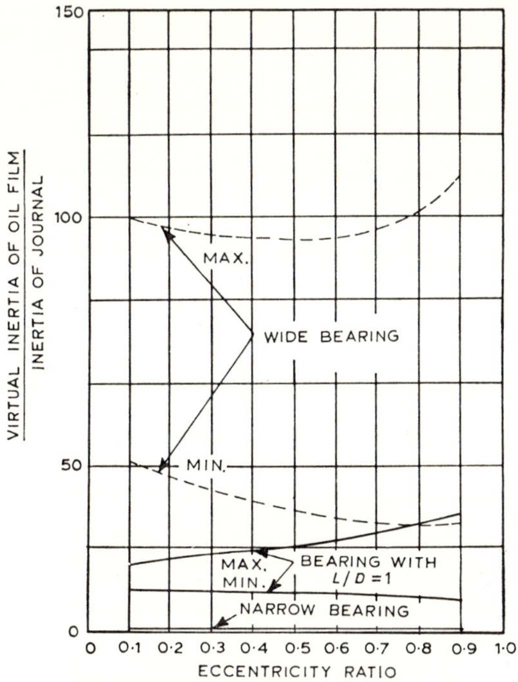

Oil lubricated bearing with steel journal.
Clearance ratio 0·001.

FIG. 10.3. Virtual inertia ratios.

method of analytical calculation of the coefficients in equations (10.1) and (10.2) with nonlaminar flow in the film. This applies to the displacement and velocity terms as well as to the acceleration terms. It may be taken, however, that so far as the oil-acceleration effects are concerned, the virtual inertia of the oil in the film is of the same order in nonlaminar flow as it would have been in laminar flow with the same journal attitude and length of film. There may be additional inertia effects due to modification of the flow regime in the film; these effects may be coupled with the displacement and velocity terms and so produce nonlinearity of the characteristics.

Summarizing, oil inertia effects bring about departure in two ways from the dynamic characteristics as conventionally calculated – by introducing acceleration terms into the characteristic relations whether the flow in the film is laminar or not, and by changing the film-flow regime above a certain speed, and thereby changing the bearing characteristics. For a bearing running in the laminar-flow regime in conditions of Sommerfeld similarity the acceleration coefficients in the linearized dynamic relations are proportional to $(Re)(C/D)$.

When the flow is laminar the inertia effects are nearly always small in their influence on rotor vibration. They may, however, be of importance with stiff light rotors carried in wide bearings.

When the film-flow regime is nonlaminar both steady-running and dynamic characteristics are influenced by oil inertia, and therefore vary with Reynolds number. It is not possible to make analytical calculations of dynamic characteristics from first principles. Those calculations which have been attempted are of an empirical nature. The very limited experimental knowledge of dynamic characteristics in nonlaminar flow regimes is reviewed in the next chapter.

CHAPTER ELEVEN

Experimental Investigation of Dynamic Characteristics

Many experimental tests have been made of the dynamic characteristics of journal bearings. The great majority of these have been tests on small-scale bearings. The results are subject to uncertainty as to the accuracy with which they reproduce dynamic conditions in bearings of much larger size. Some tests of dynamic characteristics have been made on large bearings.

Most of the tests of dynamic characteristics fall into one of the following groups.

(1) Response to disturbance excited by out-of-balance.
(2) Response to imposed reciprocating load.
(3) Investigation of liability to bearing oil whirl.

Most tests of group (2) are of interest primarily to reciprocating-engine designers. The tests confirm the 'whirl' characteristics of journal bearings – the tendency to produce motion which includes whirl in the direction of rotation even when the applied disturbance is in one direction. Reference is made later on to an instance of use of tests of this type for determination of the eight dynamic coefficients.

The tests of group (3) on bearing-oil whirl have generally taken the form of experiments on a model system of rotor and bearings. Oil whirl observed on rig tests is usually characterized by whirl frequency near half or less than half the running speed; this whirl frequency is close to the natural frequency of the rotor as affected by the characteristics of the bearing. Some types of oil whirl are sharply resonant with speed, while other types persist over a wide speed range.

Tests on bearing-oil whirl show apparently inconsistent results in the effect of oil flow, oil viscosity, and bearing loading on liability to oil whirl. These inconsistencies resemble in a general way the inconsistencies observed on turbine plant in service, but rarely give close correspondence with service experience of instability in particular cases.

The tests of group (1) with applied out-of-balance have been carried out in rigs in which observation is made of the orbit of the journal of an out-of-balance rotor. The rotor is rigid in comparison with the bearing characteristics. The disturbance forces are assessed from the inertia of the out-of-balance rotor in its orbit.

The observed orbit is normally elliptical. The measurements give four variables – amplitude and phase of movement in two coordinate directions. An out-of-balance test therefore permits of the determination of only four unknowns, as compared with the eight coefficients of the conventional linear theory of bearing dynamic characteristics.

An expression involving four coefficients gives an imperfect representation of bearing dynamic characteristics and is of no use in stability problems. Nevertheless, so much of the published information on experimental dynamic characteristics is in this form that some consideration of this approach is required.

Experimenters have commonly expressed their results as non-dimensional coefficients for equations of type:

$$P_u = K_u u + B_u\left(\frac{\dot{u}}{\omega}\right) \qquad (11.1)$$

$$P_v = K_v v + B_v\left(\frac{\dot{v}}{\omega}\right) \qquad (11.2)$$

where u, v are non-dimensional displacements measured along the principal axes of the elliptical orbit of the journal; K_u, K_v are apparent stiffness (force–displacement) coefficients; and B_u, B_v are apparent damping (force–velocity) coefficients. P_u and P_v are non-dimensional forces in the directions u, v defined in correspondence with P_x, P_y in Chapter Nine.

The apparent coefficients thus obtained are not functions of the bearing characteristics alone but are influenced also by conditions external to the bearing, even when the rotor and the bearing supports are rigid. This occurs because with unsymmetrical bearing dynamic characteristics the rotor moves in an elliptical orbit, setting up disturbance forces which are not of pure out-of-balance type.

Some of the relations affecting these apparent coefficients are given in Appendix VII. There are only three independent variables among the coefficients as expressed by (11.1), (11.2). The fourth unknown determined by observation is the angle of orientation of the principal axes of the elliptical orbit. With pure out-of-balance disturbance forces on the bearing, the relation between the coefficients is:

$$\frac{K_u}{K_v} = \frac{B_u}{B_v}. \tag{11.3}$$

When the disturbance forces are affected by the rotor motion in an elliptical orbit, and are therefore no longer of pure out-of-balance type, the relation between the coefficients is affected by the non-dimensional speed Ω. For the case where the steady load on the bearings is provided by the weight of the rotor alone, this non-dimensional speed is defined by:

$$\Omega^2 = \frac{r\omega^2}{g}$$

where r = radical clearance in bearing, ω = angular velocity of journal, and g = acceleration of gravity.

When (as is usual in bearing test rigs) the steady load on the bearing is a force W which is independent of the mass M of the rotor, Ω is defined by:

$$\Omega^2 = \frac{Mr\omega^2}{Wg}$$

With rigid rotor and rigid bearing support the relation which must be satisfied between the apparent coefficients is:

$$\frac{K_u - \Omega^2}{K_v - \Omega^2} = \frac{B_u}{B_v}. \tag{11.2}$$

The analytical expression for the apparent coefficients in terms of the eight coefficients of linear dynamic theory is complicated. It is quoted in ref. [30]. The relationships are more complicated if there is flexibility in the rotor, or the bearing supports, or both. In bearing rig tests of the usual type the apparent coefficients are little affected by variation in Ω up to $\Omega = 0{\cdot}5$, but they may be substantially affected when Ω assumes higher values.

Reports of extensive bearing rig tests under out-of-balance disturbance were published by Hagg and Sankey [31], [32]. They cover several designs of bearings of 3 in. diameter. In ref. [28] the calculated response characteristics of figure 9.12 are compared with response estimated from the nearest Hagg and Sankey test results. Agreement is good at high eccentricity ratio and fair at moderate eccentricity ratio, but poor at low eccentricity ratio. The maximum amplitude of journal motion assessed from Hagg and Sankey is always greater than that given by the calculated dynamic characteristics.

Duffin and Johnson [24] publish a set of test results of out-of-balance response on a large bearing operating in the laminar-flow regime. These test results were obtained on the equipment described in ref. [30]. The observed coefficients are compared with coefficients conventionally calculated both for the 'elliptical' bearing tested and also for a full circular bearing of the same machined bore. Reasonable agreement is found between the observed coefficients and the calculated coefficients for the test bearing over a wide range of values of duty parameter.

Few test results have been published for dynamic characteristics of bearings operating in nonlaminar-flow conditions. Observation of rotors in service has shown that there is no discontinuity in dynamic characteristics as the bearing passes into and through the nonlaminar regime. There is a tendency, in some cases at least, for stiffness and damping to increase with Reynolds number at the same duty parameter.

The few available test results suggest that dynamic characteristics are fairly close to linearity for small amplitudes. Stiffness increases with increasing amplitude, as would be expected, while damping appears little changed. These results are consistent with observations of rotor behaviour in service. Rig-test results are not available for the high amplitudes at which major nonlinearities occur.

In recent years it has been possible to measure journal attitude in bearings with accuracy, and hence to obtain force–displacement coefficients by differentiation of the observed steady-running characteristics. If the assumptions of the conventional theory hold good the force–displacement characteristics thus obtained agree with the force–displacement terms in the linearized dynamic characteristics. It must be remembered, however, that owing to uncertainties affecting these conventional assumptions, force–displacement characteristics obtained by differentiation of steady-running conditions are not necessarily consistent with the dynamic characteristics in vibration.

Experimental determination of the complete set of eight dynamic coefficients in equations (9.3) and (9.4) is a formidable problem. Out-of-balance tests require to be supplemented or replaced by other tests, with periodic disturbance forces of a type or frequency different from synchronous out-of-balance. The simultaneous equations formed from such test results may be ill-conditioned and may fail to yield satisfactory solutions unless the observations are of very high accuracy. Small non-linearities are liable to have a serious influence on the results.

Test results giving the eight coefficients for bearings of 120 mm diameter, of several designs, have been published by J. Glienecke [33]. The tests were carried out with two vibration exciters, of synchronous frequency, which imposed reciprocating load in perpendicular transverse planes. These bearings were running within the laminar regime.

The test rig for large journal bearings described in ref. [30] includes provision for application of disturbance forces at frequency independent of the running speed. The only published dynamic test results from this rig are for synchronous out-of-balance disturbance.

CHAPTER TWELVE

Influence of Bearings in Turbomachinery Vibration

Bearing characteristics affecting vibration

Journal-bearing dynamic characteristics are shown in the previous chapters to be of great complexity compared with characteristics of the simple elastic elements which turbine engineers frequently assume in vibration analysis. Some of the results established are briefly summarized below. In the bearing characteristics the disturbance forces include velocity components which are of the same order of magnitude as the displacement components, giving high damping. Additional complications arise from the nonreciprocal property of the displacement components and from the nonlinearity of the characteristics.

The bearing characteristics are not reciprocal either for small or for large amplitude of movement. Effects arising from this include easier excitation of whirl in the direction of rotation of the journal and tendency to promote instability in certain conditions of whirl.

The dynamic characteristics are approximately linear for small amplitudes of journal vibration, but they are not necessarily very close to linearity for very small amplitudes. The characteristics depart rapidly from linearity as journal movement takes up the bearing clearance. The nonlinearity is accentuated by the discontinuity in characteristics consequent on change in oil-film regime for different conditions of journal vibration.

Stability in steady running is possible only when the bearing characteristics are of restoring type, capable of locating the journal when it is vibrating near its equilibrium position. When a rotor is carried in two bearings stability in steady running of each bearing is a necessary but not a sufficient condition for the dynamic stability of the system of rotor, bearing, and bearing supports.

Rotor vibration in turbomachinery

Vibration of turbine-generator rotors usually involves flexural motion of the rotor, with transverse vibration of the journals in their bearings and transmission of the vibration through the bearing pedestals to the external components and the foundation. The vibration behaviour is determined by the characteristics of the system as a whole.

The axes and planes of reference for description of vibration are shown in figure 12.1.

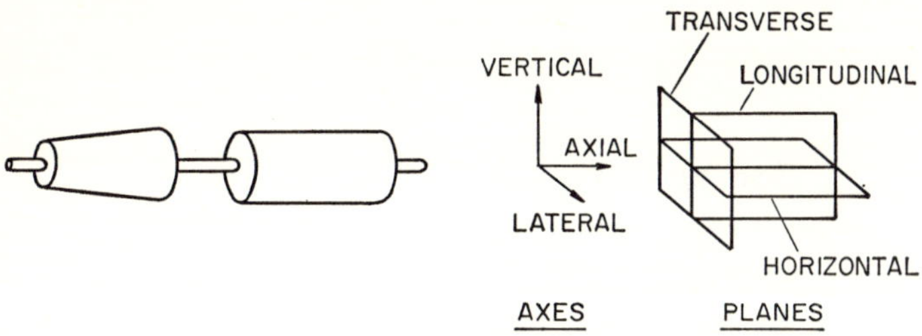

FIG. 12.1. Axes and planes of reference in vibration.

Good rotor balance, accurate alignment, and adequate margin against critical speeds have long been realized to be necessary for smooth running of turbines. Analytical and experimental work on rotor motion and on dynamic characteristics of components has been conducted on an increasing scale in recent years. Nevertheless, the understanding of rotor motion is far from complete. Progress in the development of turbine-generators has accentuated vibration phenomena which were of little importance in earlier practice [34].

Turbine vibration in which the journal bearings play a significant part falls into two classes:

(1) Vibration due to imposed disturbance – that is, response to a disturbance which is present irrespective of the state of stability of the system.

(2) Self-excited vibration – that is, vibration maintained by forces which are absent when the system is running stably, but which appear in an unstable system with frequency determined by the dynamic characteristics of the system.

If the motion of the system could be represented by linear differential equations there would be a sharp mathematical distinction between imposed vibration and self-excited vibration. The imposed vibration would be the steady motion expressed by the Particular Integral in the solution of the equations, and the self-excited vibration (if present) would be expressed by terms in the Complementary Function representing unstable motion. There are, however, non-linearities in the component characteristics, and in consequence imposed and self-excited vibration are not independent. Nevertheless, it is usually possible to distinguish whether an observed vibration is imposed or self-excited.

Vibrations of these two classes differ both in their incidence and in their treatment by the turbine design engineer. Vibration due to imposed disturbance is present in every turbine, and while the designer, manufacturer, and erector must do their best to minimize disturbance, the designer must accept the possibility of certain disturbances and must ensure that the consequent vibration is not objectionable. Self-excited vibration, on the other hand, is frequently absent, and it is the designer's objective to avoid it altogether.

Imposed disturbances are almost always periodic and associated with rotation of the rotor concerned or (more rarely) of other rotors in close proximity. Out-of-balance is the most important source of imposed disturbance; other sources include journal misalignment, unsymmetrical couplings, and periodic reactions in the generator or other driven component. With two-pole generators there are two possible sources of disturbance of frequency twice the running speed – gravitational deflection of unsymmetrical rotors and internal forces of twice the electrical frequency.

In an unstable system there is at least one component in which feedback of energy to maintain or increase vibration takes place; but this feedback can occur only when system conditions permit. In turbine plant, when instability occurs, the components providing feedback are often the journal bearings. This type of vibration is described as bearing-oil whirl.

Vibration due to imposed disturbance can usually be distinguished from self-excited vibration by the frequency. In a direct-coupled turbine line all the imposed disturbances normally present have frequencies equal to or a multiple of the running speed. When the observed vibration is of frequency equal to the running speed or an integral multiple of it the vibration is usually due to imposed disturbance.

When the vibration is of frequency less than the running speed, and there is no obvious source of imposed disturbance of this frequency, it is reasonable to conclude that the vibration is self-excited.

In imposed vibration the amplitude is diminished if it is possible to diminish the disturbance, for instance, by improving the balance or by improving the rotor alignment. In practice, continuous immunity from imposed vibration cannot be obtained in turbines because of the wander in state of balance and in state of alignment which occurs during running. This wander is due mainly to non-uniform thermal expansion.

Distortion in service of the turbine stationary parts and foundation sometimes leads to change in bearing alignment with change in journal attitude and load in various bearings. The distortion may be set up either by thermal expansion or by change in reaction transmitted through the components with varying load on set. Change in bearing load may also be set up by steam forces in turbines with partial admission.

Change in journal attitude or in bearing load affects the individual bearing dynamic characteristics, and consequently the state of vibration of the rotor.

Critical speed

Critical speed is a speed of rotation at which out-of-balance or other synchronous disturbance can excite resonant synchronous response of the rotor. Successive critical speeds correspond to different modes of vibration of the rotor in its bearings and bearing supports. Critical speed thus defined is not identical with natural frequency except when the natural frequency equals the running speed. Natural frequency varies with running speed; the causes of variation include changing bearing dynamic characteristics and gyroscopic effects.

It has long been the practice in turbine design to avoid critical speed near the running speed. Where a constant-speed turbine has in fact run near a critical speed, this has usually been due to insufficient knowledge in the design stage of the dynamic characteristics of the bearings and bearing supports. Running near a critical speed sometimes results in serious vibration, sometimes in intermittent nuisance vibration, and sometimes in slight vibration which does not give trouble.

Out-of-balance disturbance is of frequency equal to the running speed. The dynamic characteristics of the bearings and bearing supports lower the critical speeds as compared with the values which would be obtained if there were simple support at the journals. The bearings, however, limit the amplitude of vibration at critical speed, both by providing damping and by their nonlinear characteristics.

The magnitude and phase of the response to out-of-balance vary with proximity to critical speed. A simple resonant type of response is generally observed near the fundamental critical speed, and sometimes near the higher critical speeds. The vibration phenomena are, however, complicated by 'whirl' and damping effects in the bearings and by the departure from symmetry of the characteristics of both bearings and bearing supports. In some turbines complicated response occurs near the higher critical speeds, giving a band of vibration over a certain speed range with mode of vibration changing through the band, but with predominant frequency equal to running speed.

The dynamic characteristics of journal bearings and bearing supports lower the values of the critical speeds of large turbines in greater proportion at the higher critical speeds than at fundamental critical speed. The effect of flexibility of bearing supports is particularly marked on generators.

Critical speed calculation for large turbine-generator rotors has become increasingly complicated in recent years. It is standard practice to use rigidly coupled rotors instead of the former flexibly coupled rotors, so that calculation has to be made for long rotors in multiple bearings. Gyroscopic effects can often be neglected in turbine critical-speed calculation, but they are of importance with overhung rotors and in the higher critical speeds of generators. The progress of development has led to the dynamic characteristics of the bearings, supports, and foundations, playing a greater part in machine vibration. The increased analytical and experimental knowledge of bearing dynamic characteristics has, however, furnished data for calculation not previously available, while the development of digital computation has kept pace with the more elaborate calculation requirements.

The normal methods of critical-speed calculation assume linear characteristics. The only allowance for nonlinearity is made in the margin provided on the calculated critical speed. This margin also covers other uncertainties. Comparison of calculated and observed critical speeds on turbomachinery is imperative. Such comparison checks current assumptions for calculation and provides empirical information on the effect of bearings and bearing supports for approximate assessment of critical speeds on new designs.

Self-excited rotor vibration

The self-excited rotor vibrations which have been observed in turbomachinery are of frequency equal to or less than the running speed; in most cases the frequency is near or below half the running speed. These instabilities are set up by excitation in the bearings, or are permitted by the damping effect of the bearings being insufficient to restrain unstable vibration excited elsewhere in the machine.

At least three types of unstable vibration excited in the bearings have been experienced in turbomachinery:

(1) light-load instability;
(2) half-speed whirl;
(3) low-frequency whirl.

These types differ in the symptoms exhibited and in the response to remedies applied.

Light-load instability was at one time often experienced in small high-speed steam turbines, particularly when the bearing load was diminished by steam forces in partial admission. This vibration is of frequency below running speed and is often of somewhat irregular character. Light-load instability is prevented by the use of stabilized bearings which build up pressure in a groove in the top half as described by Newkirk and Taylor in 1925 [25]. A design of stabilized bearing is shown in Chapter Two. The remedy is simple and positive. There has been no occasion for turbine manufacturers to investigate light-load instability in relation to recent knowledge of bearing dynamic characteristics. Light-load instability probably occurred with a bearing running in or near the condition of instability of steady-running characteristics.

Half-speed whirl is an instability of frequency exactly half the running speed or close to half the running speed, sharply resonant in a narrow speed range. This builds up only when the natural frequency of the rotor, as influenced by the bearings, is close to half the running speed. On many rotors half-speed whirl does not occur in this condition. Liability to instability is influenced by bearing design, but no single design has been found a universal remedy. The best means of prevention is to design for avoidance of natural frequency of rotor near half the running speed. This corresponds approximately to avoiding critical speed near half the running speed.

When the speed of whirl is exactly half the running speed the whirl frequency is locked by nonlinear coupling to some vibration synchronous with speed. Whirl locked to one-third of the running speed over a limited speed range has been observed when the natural frequency was sufficiently low.

Low-frequency whirl appears over a wide speed range, with frequency little affected by the running speed. Some persistent cases have occurred with frequency well below half running speed and near to the calculated critical speed. Large generators are particularly exposed to this trouble, since it is economic to design the rotors with low critical speed ratios. This type of instability is sensitive to bearing design and operating conditions. Turbine manufacturers have evolved bearing designs which resist instability over a certain range of operating conditions. The three-land bearing design described in Chapter Two is an example.

Intermittent low-frequency whirl experienced on turbines sometimes exhibits frequency exceeding half the running speed. Intermittent instability is often associated with changes in bearing alignment which result in redistribution of bearing load and change in dynamic characteristics of the individual bearings.

In a rotor vibrating in bearing-oil whirl the exciting forces are small relatively to the inertia and restoring forces. The rotor whirls in the direction of rotation at a frequency which is generally in the near neighbourhood of a natural frequency, normally the lowest which gives substantial amplitude in a particular span. In a few cases, however, the vibration builds up with frequency and mode of deflection which suggests loss of dynamic restraint in one or more of the bearings. The stability against low-frequency whirl is controlled mainly by the balance between the non-reciprocal 'whirl' action and the damping action in the bearings, as was illustrated in 'Influence of Journal Bearings on Rotor Stability' in Chapter Nine. The bearing dynamic characteristics have an indirect influence on stability through their effect on critical speed and on mode of vibration, in addition to their direct influence on the energy interchange with a given journal motion.

Low-frequency instability excited by steam forces has occasionally been observed in turbines. In this type of instability the rotor is stable at no load, but is liable to instability at intermediate or high load. Occurrence of instability on load is, however, not a sure proof of steam excitation; it can occur through change in alignment permitting bearing-oil whirl. In a few cases steam excitation appears to have been associated with partial admission nozzles; in other cases with non-uniform circumferential distribution of steam force associated with leakage at glands or blade tips. Sometimes this type of vibration has been remedied by a bearing modification; at other times by a change in design which raised the significant natural frequency or which reduced the steam excitation.

Unstable vibration with frequency equal to running speed has occasionally been observed. The growth of amplitude is slow in this type of whirl, much slower than is normally experienced in the low-frequency types of unstable whirl. Stabilization of vibration at a moderate amplitude often takes place. So long as the set is running at constant speed, the symptoms of whirl show considerable resemblance to those due to slowly varying out-of-balance. The behaviour with changing speed is, however, different from that normally occasioned by out-of-balance. The amplitude of vibration diminishes in acceleration and increases in deceleration within a certain range of speeds; it falls off rapidly above or below the unstable speed range.

The occurrence of this type of synchronous whirl is often intermittent and apparently erratic. The consequences of the vibration vary from slight nuisance to great difficulty in starting and occasional necessity to adjust or remove the load on the set.

A critical speed can usually be recognized within the speed range liable to this type of instability. The mode of unstable vibration may be quite different from that calculated for the critical speed. Some of the phenomena of synchronous whirl can be explained by nonlinearity in the journal-bearing characteristics.

In most types of rotor instability the transition between stability and instability is influenced by a number of operating conditions, such as bearing alignment, journal alignment, oil temperature to bearings, and oil supply to bearings. These effects sometimes seem to operate in erratic and inconsistent ways. The stability boundary is determined by the net result of various actions in the system, some favouring stability and some instability.

Particular instances of instability can often be dealt with by some minor empirical modification, such as change in grooving or clearance of the bearings. It is, however, important that the basic cause of instability should be investigated, so that as far as possible risk of instability can be diminished in future designs.

All types of self-excited rotor vibration are limited in amplitude by some nonlinear effect. This nonlinear effect is generally provided by the bearing dynamic characteristics. In any instance where the bearings are ineffective the amplitude of rotor vibration may increase until damage is caused by rubbing between rotor and stator.

In spite of the number of different types and manifestations of rotor instability which have been reported, the majority of turbine generators are free from troublesome instability in service. The rare occurrence of some of the types of instability has limited opportunity for their technical investigation.

Remarks on current state of knowledge

The following general remarks apply to the current stage of knowledge on the dynamic characteristics of journal bearings in relation to the vibration of turbomachinery.

The knowledge of the dynamic characteristics of journal bearings, acquired by analysis and by experiment, has been of great value in improving the understanding of the various types of vibration of turbomachinery, and thereby facilitating the work of the turbine engineer in dealing with vibration problems. The numerical assessment of linear dynamic characteristics has been of much assistance in critical speed calculation, in spite of the gaps in the knowledge of characteristics of bearings in the nonlaminar-flow regime. Information on bearing dynamic characteristics from analysis of rig test has, however, been of limited service to turbine engineers in designing bearings for resistance to particular types of instability. Information as to how bearing design and operating conditions affect rotor stability has come mainly from observation and empirical modification of sets in service.

The difference in the value of assessed dynamic characteristics, as applied to specific problems of critical speed and to specific problems of stability of running, is associated with the greater complication of the influencing causes which determine whether or not a rotor will run stably. One complication is the balance between the various opposing effects which influence stability. Another complication is that in some types of vibration, nonlinearity plays an important part over a wide range of journal amplitudes, and in all types of vibration has a dominant influence at high journal amplitudes.

PART FOUR

Operational Experience and Prospective Development

CHAPTER THIRTEEN

Operation of Journal Bearings in Turbomachinery

Requirements in bearing performance

It is of paramount importance that bearings in turbomachinery should be reliable. Bearing failure on a large turbine generator in service has disastrous consequences. Immunity from bearing failure must cover all normal operating conditions, and also as far as possible the abnormal emergencies which occasionally arise.

Bearings must contribute to good running, especially in freedom from vibration trouble. The bearings should not require inspection more frequently than opening up of the set is required for other purposes – after the commissioning period at intervals of several years in large-steam-turbine plant. The bearings should require renewal only at long intervals to a scheduled programme.

Subject to achieving reliability and good running, the bearings should be inexpensive, easy to assemble and dismantle, low in loss, and moderate in lubricating-oil demand.

These exacting requirements result in turbine designers following conservative practices in development, with gradual and cautious adoption of new proposals in bearing design, materials, and lubricants.

Range of service conditions

Hydrodynamic journal bearings running under the designed conditions of full speed and load have a large reserve capacity. They can run steadily under far higher loads at the designed speed, and at substantially higher speeds under the designed load. This reserve capacity is obtained as a consequence of ensuring that the bearings give trouble-free running under the full range of normal service conditions. The normal range of conditions includes starting and stopping, barring, running to overspeed, and exposure to a certain amount of out-of-balance loading and other disturbances. There may be little reserve capacity at low speeds. In a few applications turbines are specified to withstand shock loading.

In most bearings in direct-coupled turbine-generators the steady load is provided by the weight of the rotor and is constant. The low speeds in starting and stopping at which this load may have to be carried in boundary lubrication, or by a hydrodynamic film of small minimum film thickness, are therefore of particular significance as regards durability of the journal and bearing surfaces. Properties of the journal and bearing materials and of the lubricant are of importance in these running conditions. In most of the larger bearings jacking oil under substantial pressure is supplied to the bottom of the bearing at low speeds, giving a local hydrostatic oil film which supports the journal independently of speed; but this does not avoid the risk of contact or boundary lubrication in emergencies.

Barring is low-speed rotation of a steam turbine or gas turbine, following operation, to ensure circumferential uniformity of rotor and stator temperatures and so to avoid thermal distortion which would cause difficulty in restarting at short notice. Barring is carried out at various speeds on different sets, from 2 to 100 r.p.m.

In steam turbines with partial admission the steam exerts transverse forces on the rotor which are vectorially added to the weight loading of the bearings. The steady load on the bearings then varies with operating condition in both magnitude and direction; this must be taken into account in bearing design.

In reduction gears the bearing load is determined largely by the tooth loading. The bearing load is often small at starting and at low speeds. This permits designing the bearings for specific loading at full speed much higher than is normal practice in direct-coupled generating plant. In some gear bearings the direction of the load swings through a large angle, and the bearing is designed to accept loading at any angle; examples have been given in Chapter Two.

The steady loads on bearings in service are readily assessed, but the dynamic loads are variable and are difficult to assess. The highest dynamic loads on turbine bearings arise from out-of-balance of a bent rotor. Liability to bending occurs particularly on the high-pressure and intermediate-pressure rotors of steam turbines. On these rotors, experience suggests that an out-of-balance load exceeding twice the load due to rotor weight is liable to be applied on at least one occasion during the service life of most rotors, and that with the more severe type of temporary or permanent bend a dynamic load of about five times the weight is liable to be applied on some rotors. It is therefore desirable for the journal bearings on these rotors to have a reserve capacity at normal speed for carrying a rotating load seven times the normal load for a few minutes without damage to the bearing.

The low-pressure rotors of steam turbines are exposed to much lower steam temperature, and rarely experience severe bending. On most rotors the rotating load due to out-of-balance is unlikely to exceed the weight load. On a few low-pressure rotors there is occurrence of emergency conditions, involving damage to the rotor, where the out-of-balance load is greater than the weight load.

On generator rotors long-term out-of-balance load is unlikely to exceed 25% of the weight, and occasional short-term out-of-balance load to exceed 80% of the weight. The exceptional occurrence of electrical failure of a coil may give a rotating load of the order of three times the weight. Some rotors are liable to bearing-oil whirl, but in consequence of the low frequency, the forces involved are rarely excessive.

On generator rotors and low-pressure steam-turbine rotors it is good practice to have bearings capable of carrying rotating load three times the normal load. Consequential damage to the plant is reduced if, in a serious emergency, the rotor can be brought to rest with undamaged bearings.

Continuity of oil supply at reasonable pressure and temperature is required for satisfactory bearing operation. The bearing can accept considerable fluctuation in oil-supply conditions. In land turbine plant with integrally driven oil pump interruption of oil supply is rare. Failures of oil supply have occurred in one or two instances on first start after erection or overhaul, due to incorrect assembly, and in other instances on shut-down, due to auxiliaries not functioning properly. Experience of marine turbine plant has shown that there is greater risk of interruption of oil supply than in land plant. Some marine bearings are designed to limit damage if the oil supply should fail.

Bearing damage and failure

In turbomachinery the failure of a journal bearing in service, necessitating shut-down and replacement, is almost unknown in conditions now recognized as good practice. Among the failures which have been experienced, some have been associated with design practices found at a later date to be undesirable. The most conspicuous of these in recent years have been the 'high chromium' failures.

The limited information on failures of journal bearings is supplemented for guidance in design by information from inspection of bearings in service, by experience on other components of turbomachinery which depend on oil-film lubrication (thrust-bearings and oil-seals), and by experience of journal bearings in plant of other types.

Damage to journal bearings can occur in a number of different ways, each of which is influenced by a variety of causes. The damage may be initiated in any normal or abnormal running condition. Influencing causes include mechanical design, construction, and support of bearing; bearing surface finish; journal surface finish; alignment; smoothness of running; properties of oil; cleanliness of oil and oil-supply conditions.

It is desirable that if wear or other surface damage takes place the damage should be confined to the bearing surface (which is comparatively readily renewed) and should not extend to the journal surface. This is one of the reasons for the almost universal choice of babbitt metal for the bearing material in turbomachinery. Journal surfaces do not experience much damage; when there is damage it is usually scoring. After a severe wipe, a journal may show blue-temper colours and occasionally heat-treatment cracks. Bearing surfaces are subject to various forms of wear, including wiping, scoring, and cracking. The position and appearance of the wear give great assistance in identifying the contributory causes of the trouble.

The wear occurs mainly with a film of very small minimum film thickness or with boundary lubrication. Local mild wear, such as characterizes running-in, is normal and acceptable. Severe wear is a symptom of trouble. The conditions leading to wear are: contact of asperities on the imperfect journal and bearing surfaces; scoring by solid particles; development of excessive local pressure or temperature; and electrical pitting.

These considerations indicate why it is of such importance to have good surface finish, good alignment of bearing and journal, freedom from local distortion of bearing, and clean well-filtered lubricant. They give some explanation of the persistence of rule-of-thumb methods in journal-bearing design, in spite of increased knowledge of the hydrodynamic phenomena in the film.

Heavy wear in barring has been observed with a combination of low speed, high load, and high bearing temperature.

Defective alignment causing vibration and wear has been found to occur with a spherically seated bearing in which the spherical seating was too tight.

Fatigue cracking of the bearing metal can be set up either by heavy vibration or by cavitation in the lubricating-oil film. Fatigue cracking has become rarer with the use of a thin layer of white metal on a steel backing. Cavitation occurs with a severely restricted oil supply; the restriction may be in the external supply, or in the bearing itself, due to the supply groove being unsuited to the journal attitude under the applied load. Cavitation gives an unpleasant noise which can be heard through a listening-rod.

Corrosion has occasionally given trouble, either by direct attack on a bearing-metal constituent or through corrosion fatigue. Corrosion has sometimes been due to oxidation products in the oil or to contamination by sea-water; but good maintenance (including regular oil changes) and the use of additives in the oil have contributed to minimizing corrosion trouble. On the other hand, cases have been reported where additives have caused trouble, either directly by corrosion or by removing paint and thereby contaminating the lubricating-oil system.

Electrical erosion of bearings and journals has sometimes given trouble. Small shaft voltages are set up by electromagnetic induction of the generator rotor. Higher voltages occasionally observed appear to be produced electrostatically, either in generator or in turbine. Bearing erosion is prevented by effective insulation of the bearings from earth in conjunction with the use of an earthing brush correctly placed on the rotor. Both insulation and brush gear are in oily environments and require good maintenance as well as good design.

Failures described as 'high-chromium type' were experienced on journal bearings and hydrogen seals in turbine-generators following the introduction of rotor steels containing about 3% chromium and 12–14% chromium. These failures occurred on a small percentage of the rotors of those materials in service, and under conditions in which journals of other steels have run successfully. The failures were characterized by local deep wear of the journal together with some general scoring. A hard black scab was formed in the white metal which acted as a cutting tool. This scab contained a high proportion of carbides. The wear debris was often in the form of steel slivers from the journal surface, giving the appearance of wire wool. Most of the failures took place in works test or in early running on site.

Failures were also experienced on marine turbine thrust and journal bearings with thrust collars or journals made from a wider range of materials, some of lower chromium content. These failures occurred with lubricating oil containing extreme-pressure additives, as used in geared sets. The failures were generally similar in appearance to the high-chromium-type failures. Both groups are covered by the alternative descriptions 'wire-wool-type' and 'machining-type' failures.

Many theories have been put forward as to the mechanism of wire-wool-type failure, and extensive analyses of experience and research tests have been made. References [35], [36], and [37] report some of this work and list earlier contributions.

The work has shown that liability to failure is influenced by the material analysis of the journal (or thrust collar) and by the additives in the oil. It is accepted that failure is initiated by foreign particles which are liable to enter the bearing even with a high standard of cleanliness of the lubricating-oil system. The resulting scoring can give rise to a self-propagating failure when the materials are susceptible. Recommendations for prevention of wire-wool-type failure are to use journal, seal, and thrust materials with less than 1·9% chromium content and with low manganese content, and to avoid the use of certain additives in the lubricating oil.

Lubricating oil

The lubricating oil used almost universally in the bearings of turbomachinery is a petroleum product which has undergone distillation, refining, and treatment with additives. The oil consists of a mixture of many chemical compounds, mostly hydrocarbons, of different structures and different molecular weights. The oil properties vary in a complicated way with the origin of the crude petroleum and the refinery treatment. A review of properties in relation to origin, treatment, and additives is given by T. I. Fowle [39]; see also Reference [57].

The British Standard Specification for steam-turbine oils, B.S. 489: 1968 (ref. [38]), covers the oils generally recommended by British manufacturers of turbomachinery. The Specification definies four grades – light, medium, heavy, and extra heavy – determined by kinematic viscosity at 100° F. Information on some oil properties (specified and unspecified) is given in Appendix VI.

Various laboratory tests are called for in the specification, but these tests do not completely ensure the suitability and durability of the oil for service in turbomachinery. Light- or medium-grade oil is generally recommended for direct-coupled turbines and heavy grade for geared turbines.

The performance of the same nominal grade of oil from the same supplier has varied over the years, in consequence of changes in the crude petroleum used, development in refinery processes, and changing use of additives. The earliest British Standard Specification (1933) called for straight hydrocarbon oil without additives. The Specification, as reissued in 1955 and 1968, called for additive materials to meet the requirements of the Standard, but did not specify the additives. Additives can give remarkable improvement in particular properties of the oil, but sometimes they have unexpected side effects.

The service life of oil in a turbine is limited by oxidation, which produces acids, sludge, and lacquer.

The rate of oxidation, and the resultant effects, vary very widely indeed with the composition of the oil and with operating conditions, such as maximum oil temperature, aeration, and materials exposed to the oil. Some materials act as catalysts, while others are liable to corrosive attack from acid formation. Anti-oxidant additives are used in turbine oils; these give extension of life and improvement of operating conditions as compared with the straight mineral oils formerly used. Anti-corrosion additives, to reduce corrosion of metallic materials, are also generally used.

Sludge of a different type can be formed in service – a water-in-oil emulsion stabilized by rust or other contaminant. The remedy for this trouble is to keep the oil system clean.

Deterioration of oil takes place at high temperature even when there is no oxygen access. In normal turbine practice there is always oxygen access, and the limiting temperature for oil oxidation, even with anti-oxidant additives, is lower than the thermal-stability limit.

In turbomachinery (and similarly in motor-car engines) the tendency over the years has been to specify and use lubricating oil of progressively lower viscosity. This development has been permitted by a higher standard of accuracy and finish in bearing surfaces, accompanied in some cases at least by the use of smaller clearance ratios in bearings of the same size and speed. The progress in refinery technique has given lubricating oil which is less erratic and more durable in service; but vigilance is necessary on the part of the operating engineer, since surprises still occur in the behaviour of batches of lubricating oil.

CHAPTER FOURTEEN

Prospects in Development of Journal Bearings

General prospects

Many proposals have been made for the future development of journal bearings, affecting principles and details of design as well as materials and lubricants. The ever-widening range of analytical and experimental research has covered aspects of these proposals. Specialized requirements, such as those of nuclear plant and of space equipment, have led to the use of novel lubricants, including liquid metals and pressurized gases.

In this chapter attention is limited to developments which may be applicable to conventional turbomachinery of moderate or substantial size, such as steam- and gas-turbine plant for land or marine application.

Evolution of current practice

In current practice in large turbine generators hydrodynamic journal bearings with lubricating oil circulated through external coolers are practically universal. Without external oil circulation the temperature rise in the bearings would be excessive.

In British and European practice the bearing surfaces are usually fixed. The alternative of tilting-pad journal bearings is widely used in American practice. Both fixed land and tilting-pad designs are susceptible to continued improvement. A bearing now in development which combines a fixed land and tilting-pads is described in ref. [24].

On the types of fluid-film bearings currently used in turbomachinery the design is such that, if wear takes place, the wear occurs on the bearing surface (rather than the journal surface) and the bearing surface is readily renewable. There is a good case for continuing this practice.

On large journal bearings with nonlaminar flow in the film the bearing loss is of serious magnitude. The loss may be reduced by reducing the bearing area. Such a reduction has to be carried out with judgement. Reduction of bearing area in the bottom half increases specific loading, and with current materials and methods of construction increases the risk of service trouble. Reduction of bearing area in the top half may affect both control of rotor vibration and reserve capacity for withstanding vibration without damage to the bearing.

If higher specific pressures are to be used special attention must be given to the detailed mechanical design of the bearing to minimize risk of local distortion of the bearing surface. For reduction of bearing loss the loss in the oil grooving must be kept low as well as the loss in the bearing oil film.

Externally pressurized bearings

Hydrostatic journal bearings, pressurized by an external supply of high-pressure fluid, have been frequently proposed with various lubricants. Gaseous lubricants could not be used in large turbine-generator plant except with pressurized bearings. As previously mentioned, pressurization by jacking oil is normal practice in the larger journal bearings for starting and barring purposes; but hydrostatic journal bearings have seldom been used at normal speed in turbine-generators.

A disadvantage of hydrostatic bearings in turbine plant is that if the bearing is to have the reserve capacity for coping with emergencies which is given by conventional hydrodynamic bearings the lubricant must be supplied at a pressure sufficient to handle the emergency loading. This requires lubricant supply at high pressure, and leads to high pumping loss, due both to increased pumping pressure and to increased oil flow.

In hydrostatic bearings for high peripheral speeds, operating with a liquid lubricant of normal density even if it is of low viscosity, the flow within the pressurized grooves is likely to be turbulent, and the loss in the grooves is likely to be in the neighbourhood of that in hydrodynamically lubricated lands of the same bearing area.

If a gas is used as the hydrostatic lubricant the turbulent loss in the grooves is kept low, but the pumping loss is increased (compared with liquid lubricant) unless very small bearing clearances are used. Very small bearing clearances give increased liability to serious running troubles.

Grooved bearings

In recent years investigation has been made of journal bearings with helical grooves, and of thrust bearings with spiral grooves, in which hydrodynamic pumping action in the grooves plays the major part in building up the bearing oil pressure [40]. Externally pressurized bearings in which the pressure distribution is influence by grooving have also received attention [41]. The pumping action in grooves has long been utilized in the functioning of journal bearings; for instance, in the stabilized bearing described in Chapter Two. Helical and spiral grooved bearings have been used in practice only to a limited extent, and only in small high-speed bearings. There is scope for investigation of the application of pressure-building grooves in the design of larger journal bearings for turbomachinery.

Flexible-element bearings

A. G. M. Michell, following his proposals for tilting-pad thrust and journal bearings (1905), designed flexible-element journal bearings in which the bearing pads were integral with a carrier ring but were flexibly mounted so as to tilt under load. Flexible-element journal bearings were manufactured and put into service, but were not widely adopted.

In more recent years A. Tondl has carried out extensive tests on the stability of model rotors running in journal bearings of various types; these types included several designs of flexible-element bearings [42]. The flexible-element bearings showed a wide range of resistance to instability. Certain designs gave stable running over the full range of conditions tested, with running speed up to eight times the fundamental critical speed. Such a design might be of service in cases of vibration trouble.

The use of flexibly mounted bearings (as distinct from flexible bearing pads) has been frequently proposed for the purpose of adjusting critical speeds, and sets incorporating this proposal are in service.

Ball and roller bearings

Ball and roller bearings are universally used in aero gas turbines. The bearings used have been specially developed for this application.

Rolling contact bearings have been little used as main bearings in turbomachinery for land or marine service, except in gas-turbine plant adapted from jet engines. The use of ball and roller bearings in aero gas turbines eases starting and permits of lubricating-oil systems which are smaller and lighter than those for plain bearings. Aero gas turbines require strip inspection at fixed overhaul periods of much shorter duration than the periods normal in steam-turbine plant, and the ball and roller bearings are usually renewed on these inspections. The various types of failure to which rolling contact bearings are liable are discussed in papers to the 1967 I.Mech.E. Lubrication and Wear Conference [4].

The durability of ball and roller bearings as applied in aero gas turbines does not approach the durability obtained in plain journal bearings in steam turbines. Ball and roller bearings have not been developed for the size, load, and speed requirements of large land turbine plant. Such development would be very difficult. The required operating conditions are well outside the range recommended by M. J. Neale for rolling contact bearings [43].

Materials

The material of the journal in turbomachinery is usually the steel of the shaft or rotor. Since the 'high-chromium' type of failure has been recognized, rotors of high chromium content have been avoided or have been designed with sleeved journals. The other rotor materials in general use have been found acceptable for service as journals with oil having no extreme-pressure additive. The 'high-chromium' experience shows that precautions must be taken in the introduction of new journal materials.

Selection of the bearing material is of major importance. A bearing material is required which is readily 'run-in' and which is unlikely to damage the journal if rubbing contact should take place either directly or through particles carried in the lubricant. This calls for a soft bearing material with the steel journal. The ideal bearing material should possess a whole series of properties, not all of which can be explicitly defined; the problem is reviewed in ref. [44]. In choosing the material a compromise has to be made between the desired properties. Laboratory tests have been devised for some of these properties, but the final choice must be based on experience in service.

Tin-base babbitt is generally used as the journal bearing material in turbomachinery. A material widely used by British manufacturers is No. 2 alloy of B.S. 3332 : 1961. This alloy is used regularly with nominal loading up to 250 lb/sq in. in journal bearings which start under load. Higher loading at full speed has been used in bearings with low loading at low speed, also in a few cases of large sets with bearings starting under load.

Thinning the babbitt lining on a steel backing leads to increased fatigue strength, but this thinning increases the manufacturing problems and diminishes the reserve for wear.

Consideration has been given to various copper–lead, lead–bronze, and aluminium alloys which have better fatigue and temperature properties than babbitt, but are inferior in wear and boundary lubrication properties. A number of these materials have been used under heavy loading in the bearings of reciprocating engines, usually as a thin lining on steel, sometimes as a composite lining of two alloys on steel. A similar combination of materials might be used in development of a journal bearing for high-pressure, high-temperature service in turbomachinery.

Plastic bearing materials, often lubricated by water, have long been used in heavily loaded, low-speed bearings, as in rolling mills. Development is being carried out in plastic bearing linings for higher speeds, but up to the present these plastic bearings have not been used in turbomachinery.

Much work has been done on the development of self-lubricated bearing materials for certain applications [45]. In conventional turbomachinery the combination of load and speed leads to excessive heat generation, even with the lowest friction coefficient which such materials can give; fluid film lubrication with circulated lubricant is imperative. Self-lubricated bearing materials are of no advantage in turbomachinery.

Lubricants

The development of mineral oils for turbine lubrication has been carried out mainly by the oil manufacturers. Development continues in this field, both in refinery processes and in treatment by additives, with the objects of improving the product, increasing service life, and producing lubricants suited to more onerous operating conditions [39]. Much development work is in progress on synthetic oils. This work is already applied in lubrication of aero gas turbines and may later influence practice in turbomachinery with film-lubricated bearings.

Laboratory tests and experience in other applications give some forecast of the behaviour of a lubricant in turbine service, but the ultimate check must always come from operation.

Experience shows that it is imperative for a lubricant in turbomachinery to have reasonable properties in boundary lubrication, even in cases where the journal when rotating is normally separated from the bearing by a film which is pressurized at low speeds. Adequate life of the lubricant in service must be obtained. The lubricant must be free from any tendency to attack the materials (including paint and insulation) to which it may have access at any time. Where circumstances justify it, plant may be designed to suit a novel lubricant by avoiding the use of material liable to attack.

Many unconventional lubricants have been proposed for turbomachinery with film-lubricated bearings. The principal objectives put forward for these lubricants have been to permit of bearing operation at higher temperature, to simplify sealing arrangements in turbines, and to reduce fire risk.

The existence of fire risk has been shown by occasional serious fires involving lubricating oil. In the usual turbine installation the same lubricating system supplies both the bearings and the relay gear operating the control valves. With steam at high pressure and temperature there is increased risk of ignition of leakage oil from components of the control gear. There are a few turbines in service in which the relay gear system is separate from the bearing lubricating system and uses a different and more fireproof fluid medium.

The unconventional lubricants which receive mention here are synthetic oils, water, water-containing lubricants, steam, air, and gas.

Synthetic oils

A wide range of synthetic oils have been developed in a number of chemical groups [39] and [46]. All these oils are much more expensive than the ordinary turbine lubricating oils; the cheapest synthetics are about five times the cost of mineral lubricating oil, and some synthetics are far more expensive. A synthetic oil will therefore be used in turbomachinery only if it offers outstanding advantage in service.

Some of the synthetic oils have been developed for specialist lubrication applications in other fields. The application of greatest interest in the context of turbomachinery is to lubrication of aero gas turbines for supersonic aircraft [47]. These turbines have rolling contact bearings;

the lubricant is circulated, but heat can be removed only at a fairly high temperature; short life is acceptable for the lubricant. Operating conditions are therefore widely different from those in turbomachinery for land or marine application. In the aero-gas-turbine application synthetic oils have been used at bulk operating temperatures within the range 250°–300° F and are in course of development for still higher temperatures. These are far higher than the temperatures for which long life can be expected from mineral lubricating oils exposed to air. Synthetic oils from at least four different chemical groups have been used. At the high operating temperatures contact of some of the oils with certain metals has to be avoided.

Synthetic oils are available in a considerable range of viscosities. The rate of variation of viscosity with temperature is generally less than for mineral oil; in some cases much less. Synthetic oils are higher in density than mineral oil; most of them are only slightly higher, but one type is more than double the density of mineral oil.

If the need should arise in land turbomachinery for operation of bearings at temperature much higher than at present, and there is economic justification for the special development involved, the experience of high-temperature lubrication in aero gas turbines will provide a useful starting-point for developing the land turbine application of similar oils.

Such a development is not at present in hand for land turbines. The use of synthetic oil in land turbines has, however, been proposed for fire resistance. The synthetic oils are not completely fireproof, and in certain instances give toxic products on exposure to heat, but some of them are much more fire resistant than ordinary turbine oil. Synthetic oils have been used as fire-resistant hydraulic fluids. A few turbine relay systems have been operated on synthetic oil with varying degrees of success.

A. C. M. Wilson quotes some service experience in relay systems and reviews the properties of synthetic fire-resistant fluids for turbine application [48]. He concludes that the physical properties of the phosphate esters are advantageous for turbine application, but that chemical stability and corrosion properties render the chlorinated diphenyls preferable.

Some small steam turbines have been experimentally lubricated with fire-resistant synthetics, and unattended gas-turbine compressor units have gone into service thus lubricated. There is little published experience of the use of fire-resisting synthetic oil as lubricant in a large steam turbine.

Water

Water has long been used as lubricant in some applications, such as the large slow-speed bearings of rolling mills. It has more recently been used in the high-speed, lightly loaded bearings of feed pumps. Its advantages in this application include simplification of sealing arrangements, freedom from contamination of the feed by the lubricant, and completely fireproof qualities. Corresponding advantages could be obtained by the use of water in steam-turbine bearings. Disadvantages arise from low viscosity, poor properties in boundary lubrication, and tendency to promote corrosion.

In large turbine bearings, with nonlaminar flow in oil or water, the loss with either lubricant would be in the same neighbourhood. The high specific heat of water (more than double that of oil) assists in limiting temperature rise through the bearing.

It has been proposed that feed water should be used as lubricant for recovery of heat from bearing loss. This would be convenient with simplified sealing arrangements between steam and water.

Extensive rig tests on water-lubricated bearings have been carried out, particularly for starting and low-speed conditions. The tests by Hother-Lushington and his colleagues reported in refs. [49], [50], and [51] include much work on journal and bearing materials and on their surface finish. Some interesting results on hard bearing materials are reported.

There is no doubt that water could be used as a lubricant in large steam turbines, but continued development would be required to ensure success.

With water as lubricant, hydrostatic bearings have been suggested as an alternative to hydrodynamic bearings. Hydrostatic bearings do not seem necessary (except for jacking pressure in starting) with a liquid lubricant which runs in a nonlaminar film regime in a hydrodynamic bearing. The minimum film thickness in a water-lubricated turbine bearing at full speed will be approximately the same as in the corresponding oil-lubricated bearing.

Water-containing lubricants

A water-containing lubricant offers the possibility of combining fire resistance with other properties which permit of approach to conventional journal bearing practice. Normal sealing arrangements for the lubricant must be retained.

Tests with water-base lubricant were reported by F. Samuelson in 1937 [52]. The lubricant was water containing 2% soluble oil; the soluble oil was a cutting fluid for machine tools. Service tests covered thirty months' satisfactory running on a 25-kW steam turbine. Standard designs and materials were used in the bearings; the clearances in the oil pump and the governor oilways were diminished. The peripheral speeds and loads on the bearings of this set were low. Rig tests at higher loads and speeds were carried out on bearings of 6 in. diameter up to 6,000 r.p.m.

Hydraulic fluids consisting of emulsions of oil in water, emulsions of water in oil, and water–glycol solutions have undergone much development since Samuelson's tests [39]. The fluids of high water content have a high degree of fire resistance. With most of these fluids it is desirable to limit continuous operating temperature to about 65° C. to ensure stability. The fluids are far superior to water in corrosion characteristics, and have reasonable properties in boundary lubrication. Some of them promote fatigue failure in heavily loaded ball and roller bearings. If the need should arise, development tests could be used to select an existing water-containing lubricant which could be applied in large turbine plant with little change to the lubricating system.

Steam, air, and gas

Air and gas bearings, both hydrodynamic and pressurized, have been used in light high-speed turbomachinery for special purposes. Troubles have been experienced, particularly in starting and in unstable whirl, but success has been achieved within a certain range of applications. Numerous designs of gas bearings have been used or proposed. Many analytical and experimental investigations on gas bearings have been published [53].

The application of gaseous lubricants in the journal bearings of heavy turbine plant presents far greater problems. Proposals have, however, been made for steam bearings in steam turbines, for air bearings in gas turbines, and for air or gas bearings in generators. Some preliminary experimental work for these applications has been carried out.

Any one of these applications will require extensive development in bearing design, in plant design, and in experimental work. In heavy plant pressurized lubrication with hydrostatic bearing-films will be required in all cases at starting, and in most cases at all running speeds.

Continuous supply of pressurized lubricant at the main bearing films is essential, otherwise there would be immediate bearing failure. It may be possible in some cases to supplement external pressurization by hydrodynamic action in the bearing at speed.

The question of reserve capacity of the bearings is important. With externally pressurized bearings the reserve capacity must be provided by design; it is not obtained as a bonus, as occurs in oil-lubricated bearings of conventional type.

The advantages sought from steam or air lubrication in turbines include simplification of sealing arrangements, reduction of gland leakage, reduction of fire risk, and ability to operate the bearing at higher temperature. Most of these advantages could only be attained by radical redesign of the turbine plant, as compared with present designs which have been evolved by experience.

In the case of steam plant, steam at adequate pressure for the bearings can often be obtained from the boiler. Control gear for steam temperature at least will be required in the steam supply, to limit thermal distortion.

In the case of gas-turbine plant, air from compressor outlet is not available at starting, and at full speed the pressure may be inadequate for hydrostatic journal bearings. Auxiliary air-compressor plant will have to be provided.

On generators it has been suggested that the coil retaining-rings at the ends of the rotor body, or alternatively, the whole rotor body, should be gas-borne, replacing separate journal bearings. The lubricant would be the same gas as used for reducing windage loss – in current practice, hydrogen. Owing to the large bearing area which could be provided, moderate gas pressure would be sufficient.

This revolutionary proposal would involve major changes in the whole design of the generator, including mechanical structure and cooling arrangements. The advantages in prospect are shortening of the machine and easing of the deflection and critical speed problems on a long rotor.

It is clear that all the proposals for gas-lubricated bearings in heavy turbine-generator plant are of long-range type, and could not be put into service without several years' extensive development.

Appendices

APPENDIX I

Analytical Solution for Very Narrow Bearing

Steady running

The Ocvirk approximation for the narrow bearing case [7] is based on the assumption that when L/D is sufficiently small the circumferential pressure gradient in the film is everywhere of smaller order than the axial pressure gradient. This permits the first term to be dropped from the left-hand side of equation (4.3), leaving

$$\left(\frac{R}{L}\right)^2 \frac{\partial}{\partial z}\left\{(1 + n \cos \beta)^3 \frac{\partial p}{\partial z}\right\} = -12\pi n \sin \beta$$

Hence
$$\frac{\partial^2 p}{\partial z^2} = -12\pi\left(\frac{L}{R}\right)^2 \frac{(1 + n \cos \beta)^3}{n \sin \beta}.$$

At the sides of the bearing, $p = 0$ at $z = -\frac{1}{2}$ and $+\frac{1}{2}$. Integrating and rearranging,

$$p = 6\pi\left(\frac{L}{R}\right)^2\left(\frac{1}{4} - z^2\right)\frac{n \sin \beta}{(1 + n \cos \beta)^3}. \quad \text{(I.1)}$$

For an assigned value of eccentricity ratio n, p is thus given as a function of the coordinates z and β.

Over the bearing surface, $z = -\frac{1}{2}$ to $z = +\frac{1}{2}$, p is positive from $\beta = 0$ to $\beta = \pi$, and negative from $\beta = \pi$ to $\beta = 2\pi$. The pressure-bearing film is taken as extending over the arc with p positive, and $p = 0$ is assumed over the other part of the bearing surface.

The form of (I.1) shows that for the narrow bearing $p(R/L)^2$ gives a measure of dimensionless pressure which is independent of the width ratio of the bearing. Hence the parameter which determines the operating condition of the narrow bearing may be taken as

$$S\left(\frac{L}{D}\right)^2 = \frac{\mu DLN}{W}\left(\frac{L}{C}\right)^2.$$

This is defined by Ocvirk as C_n, the capacity number of the narrow bearing.

Evaluation of the integrals and calculation as in Chapter Four leads to the results

$$\tan \phi = \frac{\pi(1 - n^2)^{\frac{1}{2}}}{4n} \tag{I.2}$$

$$C_n = S\left(\frac{L}{D}\right)^2 = \frac{(1 - n^2)^2}{\pi n[\pi^2(1 - n^2) + 16n^2]^{\frac{1}{2}}}. \tag{I.3}$$

Linear dynamic characteristics

The following analytical expressions for the linear dynamic coefficients in (x, y) coordinates are obtained by the principles of calculation described in Chapter Nine.

$$a_{11} = \frac{4\{\pi^2(2 - n^2) + 16n^2\}}{[\pi^2(1 - n^2) + 16n^2]^{\frac{3}{2}}}$$

$$a_{22} = \frac{4\{\pi^2(1 - n^2)(1 + 2n^2) + 32n^2(1 - n^2)\}}{(1 - n^2)[\pi^2(1 - n^2) + 16n^2]^{\frac{3}{2}}}$$

$$a_{12} = \frac{\pi\{\pi^2(1 - n^2)^2 - 16n^4\}}{n(1 - n^2)^{\frac{1}{2}}[\pi^2(1 - n^2) + 16n^2]^{\frac{3}{2}}}$$

$$a_{21} = -\frac{\pi\{\pi^2(1 - n^2)(1 + 2n^2) + 32n^2(1 + n^2)\}}{n(1 - n^2)^{\frac{1}{2}}[\pi^2(1 - n^2) + 16n^2]^{\frac{3}{2}}}$$

$$b_{11} = \frac{2\pi(1 - n^2)^{\frac{1}{2}}\{\pi^2(1 + 2n^2) - 16n^2\}}{n[\pi^2(1 - n^2) + 16n^2]^{\frac{3}{2}}}$$

$$b_{22} = \frac{2\pi\{n^2(1 - n^2)^2 + 48n^2\}}{n(1 - n^2)^{\frac{1}{2}}[\pi^2(1 - n^2) + 16n^2]^{\frac{3}{2}}}$$

$$b_{12} = b_{21} = -\frac{8\{\pi^2(1 + 2n^2) - 16n^2\}}{[\pi^2(1 - n^2) + 16n^2]^{\frac{3}{2}}}.$$

APPENDIX II

Analytical Solution for Infinitely Wide Bearing

Steady running

The analytical solution for the infinitely wide bearing of circular bore, with the conventional assumptions, is outlined below.

For the very wide bearing, conditions are uniform axially. $\partial p/\partial z = 0$ and the differential equation (4.3) becomes

$$\frac{\partial}{\partial \beta}\left\{(1 + n \cos \beta)^3 \frac{\partial p}{\partial \beta}\right\} = -12\pi\, n \sin \beta. \qquad \text{(II.1)}$$

The pressure-bearing film extends from $\beta = 0$ to $\beta = \beta_2$. $p = 0$ at each end of the film; β_2 is determined by the additional condition that $\partial p/\partial \beta = 0$.

The integrations are made using the Sommerfeld substitution

$$\cos \psi = \frac{n + \cos \beta}{1 + n \cos \beta}. \qquad \text{(II.2)}$$

It is found that ψ_2, the value of ψ corresponding to β_2, is given by

$$n(\psi_2 - \sin \psi_2 \cos \psi_2) = 2(\sin \psi_2 - \psi_2 \cos \psi_2). \qquad \text{(II.3)}$$

The results of the calculation are

$$\tan \phi = \frac{(1 - n^2)^{\frac{1}{2}}(\psi_2 - \sin \psi_2 \cos \psi_2)}{(1 - \cos \psi_2)^2} \qquad \text{(II.4)}$$

$$S = \frac{(1 - n^2)(1 - n \cos \psi_2)}{3\pi n[(1 - \cos \psi_2)^4 + (1 - n^2)(\psi_2 - \sin \psi_2 \cos \psi_2)^2]^{\frac{1}{2}}}. \qquad \text{(II.5)}$$

These expressions involve the auxiliary angle ψ_2, and are more cumbersome than the corresponding expressions for the narrow bearing case. Their greater complication is related to the circumferential length of the film in the case of the very wide bearing being a function of the eccentricity ratio.

Linear dynamic characteristics

Analytical expressions for the linear dynamic coefficients are of considerable complication. The calculations for the graphs plotted were made from the steady-running characteristics using the relations given in Chapter Nine. The formula for the coefficient b_{kk}, which cannot be calculated from the steady-running characteristics, is

$$b_{kk} = \frac{3\pi S}{(1 - n^2)^{\frac{3}{2}}}\left\{2(\psi_2 - \sin\psi_2\cos\psi_2) - \frac{[2(1 - \cos\psi_2) - n\sin^2\psi_2]^2}{(\psi_2 - n\sin\psi_2)(1 - n\cos\psi_2)}\right\}.$$

APPENDIX III

Transformation of Axes

Formulae are quoted for transformation of the expressions for linearized dynamic characteristics to axes x', y' inclined at orientation θ (Fig. III.1). These formulae are applied to demonstrate some properties of the dynamic characteristics. Methods of systematic computation are not discussed.

In formulae (9.3), (9.4) for force–displacement–velocity characteristics, and in formulae (10.1), (10.2) for force–displacement–velocity–acceleration characteristics, the assumption of linearity permits the transformation of the displacement, velocity, and acceleration terms to be made independently of one another. Attention is therefore limited in the first instance to the force–displacement relations:

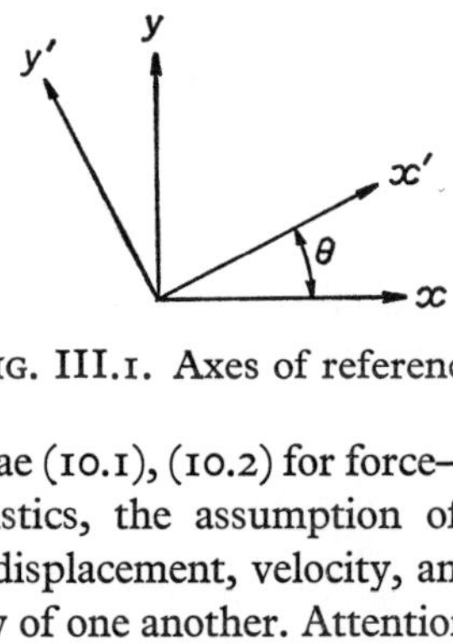

FIG. III.1. Axes of reference.

$$P_x = a_{11}x + a_{12}y \quad \text{(III.1)}$$

$$P_y = a_{21}x + a_{22}y. \quad \text{(III.2)}$$

These relations are to be transformed to

$$P'_x = a'_{11}x' + a'_{12}y' \quad \text{(III.3)}$$

$$P'_y = a'_{21}x' + a'_{22}y'. \quad \text{(III.4)}$$

in the axes inclined at orientation θ.

Calculation gives

$$a'_{11} = \tfrac{1}{2}\{a_{11} + a_{22}) + (a_{11} - a_{22}) \cos 2\theta + (a_{12} + a_{21}) \sin 2\theta\} \quad \text{(III.5)}$$

$$a'_{12} = \tfrac{1}{2}\{(-a_{11} + a_{22}) \sin 2\theta + (a_{12} - a_{21}) + (a_{12} + a_{21}) \cos 2\theta\} \quad \text{(III.6)}$$

$$a'_{21} = \tfrac{1}{2}\{(-a_{11} + a_{22}) \sin 2\theta - (a_{12} - a_{21}) + (a_{12} + a_{21}) \cos 2\theta\} \quad \text{(III.7)}$$

$$a'_{22} = \tfrac{1}{2}\{(a_{11} + a_{22}) - (a_{11} - a_{22}) \cos 2\theta - (a_{12} + a_{21}) \sin 2\theta\}. \quad \text{(III.8)}$$

The relations show that $(a'_{11} + a'_{22})$, $(a'_{12} - a'_{21})$ and $(a'_{11}a'_{22} - a'_{12}a'_{21})$ are invariant, there is, are the same for every angle of orientation. These invariant characteristics are of special significance in problems of rotor dynamics.

The directional properties of the coefficients are conveniently shown by expressing them in terms of a_r and ϕ_a, where a_r is positive and

$$a_r^2 = \left(\frac{a_{12} + a_{21}}{2}\right)^2 + \left(\frac{a_{11} - a_{22}}{2}\right)^2$$

$$\sin 2\phi_a = \frac{a_{12} + a_{21}}{2a_r}.$$

Then

$$a'_{11} = \tfrac{1}{2}(a_{11} + a_{22}) + a_r \cos(\theta - \phi_a) \quad \text{(III.9)}$$

$$a'_{12} = \tfrac{1}{2}(a_{12} - a_{21}) - a_r \sin 2(\theta - \phi_a) \quad \text{(III.10)}$$

$$a'_{21} = \tfrac{1}{2}(a_{12} - a_{21}) - a_r \sin 2(\theta - \phi_a) \quad \text{(III.11)}$$

$$a'_{22} = \tfrac{1}{2}(a_{11} + a_{22}) - a_r \cos 2(\theta - \phi_a). \quad \text{(III.12)}$$

With varying orientation of axes of reference, the direct coefficients a'_{11}, a'_{22} pass through maxima and minima at

$$\theta = \phi_a, \phi_a + \frac{\pi}{2}, \phi_a + \pi, \phi_a + \frac{3\pi}{2}.$$

The maximum and minimum values of a'_{11}, a'_{22} are

$$\tfrac{1}{2}(a_{11} + a_{22}) \pm a_r.$$

With a_{11}, a_{22} both positive, the whole range of values of a'_{11}, a'_{22} is positive, provided that

$$\left\{a_{11}a_{22} - \left(\frac{a_{12} + a_{21}}{2}\right)^2\right\}$$

is positive.

When the force–displacement system is reciprocal $a_{12} = a_{21}$ and the axes of reference which give maximum and minimum values of a'_{11}, a'_{22} are the principal axes which make

$$P'_x = a'_{11}x'$$

$$P'_y = a'_{22}y'.$$

In general, however, the force–displacement systems in journal bearings are non-reciprocal and $(a_{12} - a_{21})$ is positive, giving whirl effects in direction of journal rotation. In such a case there are no principal axes. In the orientations $\theta = \phi_a$, etc., which give maximum and minimum values of a'_{11} and a'_{22}, the relation $a'_{12} = -a'_{21}$ holds.

The transformation of the force–velocity relations is carried out by formulae similar to those given for the force–displacement relations, replacing 'a' everywhere in the formulae by 'b'. Conventional theory gives reciprocal force–velocity characteristics, making $b_{12} = b_{21}$. The value of ϕ_b is not the same as of ϕ_a. When the bearing characteristics are strongly unsymmetrical (as happens at high eccentricity ratio) there is a tendency for ϕ_a and ϕ_b to approach one another.

The transformation of the force–acceleration relations is carried out in the same way, replacing 'a' everywhere in the formulae by 'c'. The conventional theory gives reciprocal force–acceleration characteristics, with $c_{12} = c_{21}$. This justifies expressing the virtual inertia characteristics of the lubricant in terms of principal axes of inertia.

The complete expressions for P'_x, P'_y are formed by adding the displacement, velocity, and (where appropriate) acceleration terms as derived above.

APPENDIX IV

Limiting Values of Characteristics at High Eccentricity Ratio

$$m = 1 - n = \frac{\text{minimum film thickness}}{\text{radial clearance}}$$

ϕ = attitude angle

m and ϕ are small.

Characteristic	*Wide bearing*	*Narrow bearing*
S = duty parameter	$0{\cdot}130m$	—
$C_n = S(L/D)^2$ = Ocvirk capacity number	—	$0{\cdot}318m^2$
ϕ	$1{\cdot}88\sqrt{m}$	$1{\cdot}11\sqrt{m}$
a_{kk}	$1/m$	$2/m$
a_{kl}	$1{\cdot}88\sqrt{m}$	$1{\cdot}11\sqrt{m}$
a_{lk}	$-0{\cdot}94\sqrt{m}$	$-1{\cdot}665\sqrt{m}$
a_{ll}	1	1
b_{kk}	$2{\cdot}68\sqrt{m}$	$3{\cdot}33\sqrt{m}$
$b_{kl} = b_{lk}$	-2	-2
b_{ll}	$3{\cdot}76\sqrt{m}$	$2{\cdot}22\sqrt{m}$

APPENDIX V

Definitions and Units of Viscosity

The viscosity is defined for steady laminar flow by

$$\mu = \frac{\text{Shearing stress}}{\text{Rate of shear}}.$$

This viscosity of lubricant is used in the analysis of this book. It is sometimes called the absolute or dynamic viscosity, for clearer distinction from the kinematic viscosity

$$\nu = \frac{\mu}{\rho}$$

where ρ = density of lubricant.

Both viscosity μ and kinematic viscosity ν have been extensively used in engineering calculations on bearings. The viscosity is the more convenient to use in calculations of bearing characteristics in laminar flow, where density of the lubricant can be ignored. The density of the lubricant enters into calculations of nonlaminar flow and inertia effects. Some of the formulae involved in these calculations are more concisely expressed in terms of ν.

Commercial measurements of oil viscosity have usually been made by viscometers which have given rise in the past to a number of arbitrarily defined measurements of viscosity. In recent years dimensional units which are precisely defined have been increasingly used. Even with these, however, care is required to avoid confusion.

Commercial viscometers generally give measurements which are approximate functions of kinematic viscosity, and hence kinematic viscosity is widely used in specifying lubricating oils. The British Standard Specification for Steam Turbine Oils specifies kinematic viscosity in centistokes.

With fundamental dimensions length L, mass M, time T, the dimensions of the derived units are:

Viscosity μ, $L^{-1}MT$

Kinematic viscosity ν, L^2T^{-1}

In c.g.s. units the unit of viscosity is termed a poise. As this unit is large compared with the viscosity of many fluids, it is customary to use the centipoise, which is one-hundredth of a poise. The c.g.s. unit of kinematic viscosity is the stoke, and the customary unit, the centistoke, is one-hundredth of a stoke. Water at 20° C has a viscosity of approximately one centipoise and a kinematic viscosity of approximately one centistoke.

The name reyn has been given to the unit of viscosity which corresponds to fundamental dimensions length, force, and time measured in units of seconds, pounds force, and inches. One reyn equals $6{\cdot}9 \times 10^6$ centipoises.

To convert to inch–lb–sec units with fundamental dimensions of length, mass, and time, viscosity in centipoises must be multiplied by $5{\cdot}6 \times 10^{-5}$, and kinematic viscosity in centistokes must be multiplied by $1{\cdot}55 \times 10^{-3}$.

In the S.I. System, with basic units metres, kilograms, and seconds in length, mass, and time, one centipoise equals 10^{-3} S.I. unit of absolute viscosity, and one centistoke equals 10^{-6} S.I. unit of kinematic viscosity.

The analytical relations of this book apply with any consistent system of units.

APPENDIX VI

Properties of Turbine Lubricating Oils

Figure VI.1 shows the viscosity range of steam turbine oils as specified by B.S. 489 : 1968 [38]. The kinematic viscosity is plotted in centistokes against temperature in ° F – the units used in the Specification. Kinematic viscosity is plotted on a log scale. Minimum and maximum viscosities are shown for each of the four grades of oil – light, medium, heavy, and extra heavy. The viscosity at 100° F defines the grade of oil. The specification quotes maximum viscosity at 70° F and maximum and minimum viscosity at 140° F for each grade; these figures are not mandatory, but are given for guidance as to the limits likely to be obtained with oils meeting the other requirements of the specification. The curves of the figure are drawn through the various points quoted.

This specification of 1968 has gaps between the viscosities specified for the various grades of oil. In the earlier issues of B.S. 489 : (1933 and 1955) there were no gaps in viscosity between the specified grades.

Densities of oil are not mentioned in the Specification and there is no explicit specification of absolute viscosity. In practice, however, the density at 70° F of oils meeting the Specification is usually within the range 0·87–0·90 grams per cc. Density falls slowly with rising temperature.

The viscosity of turbine oil varies little with pressure up to pressures of a few hundred pounds per square inch. There is substantial rise of viscosity at much higher pressures.

The specific heat of turbine oil at 100° F is about 0·47. It increases slowly with temperature.

In figure VI.2 the kinematic viscosity of the oil used in the tests discussed in Chapter Six is compared with the figures from the 1955 Standard for medium-grade oil. The viscosity is in the lower part of the range specified in 1955 for medium-grade oil, and is within the range of the 1968 Standard.

Information on other properties of lubricants, and on the properties of a wider range of lubricants, is given in refs. [54], [55], and [56].

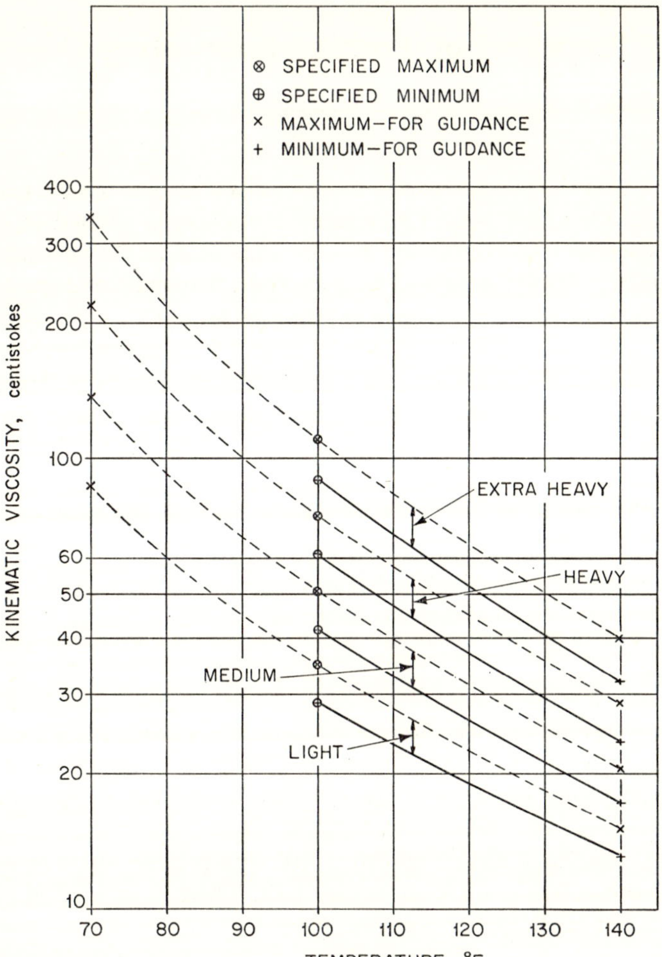

FIG. VI.1. Kinematic viscosity range of standard steam turbine oils B.S. 489 : 1968.

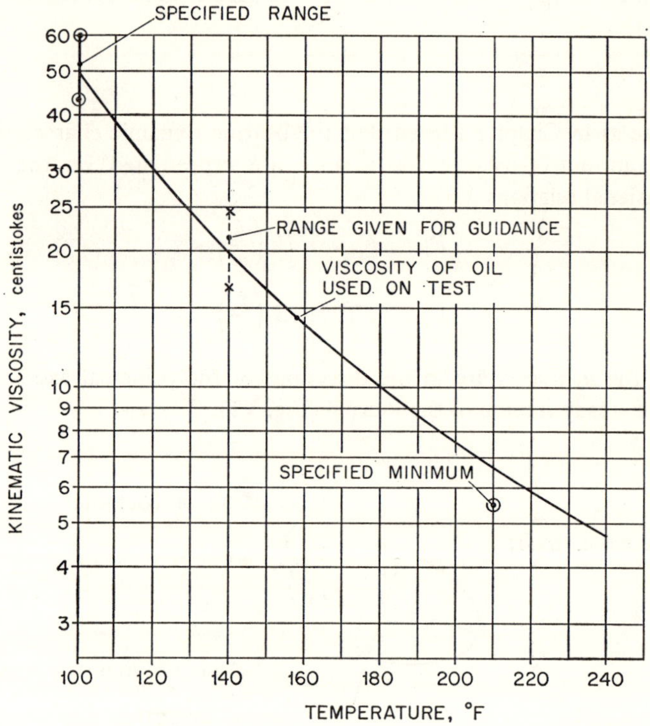

FIG. VI.2. Comparison of test oil with standard specification.

APPENDIX VII

The Apparent Out-of-balance Coefficients

Assume as in Chapter Eleven that the bearing dynamic characteristics under an out-of-balance disturbance are represented by the non-dimensional relations

$$P_u = K_u u + B_u\left(\frac{\dot{u}}{\omega}\right)$$

$$P_v = K_v v + B_v\left(\frac{\dot{v}}{\omega}\right)$$

where the u, v axes are in the directions of the principal axes of the orbit of steady motion of the journal (Fig. VII.1).

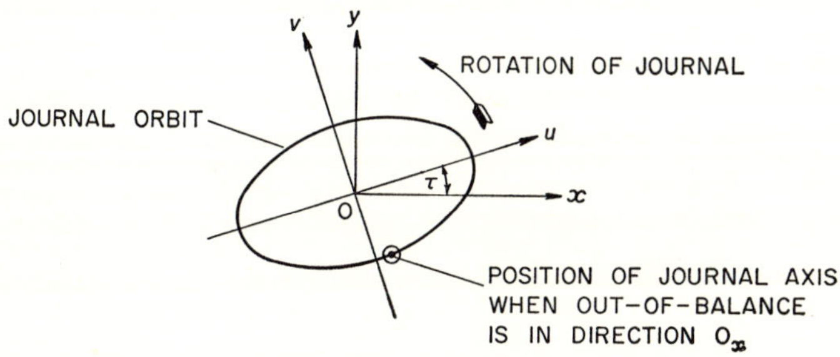

FIG. VII.1. Motion of journal axis relative to bearing.

Pure out-of-balance disturbance is represented by

$$P_u = A \cos \omega t$$

$$P_v = A \sin \omega t$$

and the harmonic response to the disturbance is represented by

$$u = C \cos (\omega t - \delta)$$

$$v = D \sin (\omega t - \delta)$$

where A, C, D, δ are constants.

By substitution, it is found that the above relations can be satisfied only when

$$\frac{K_u}{K_v} = \frac{B_u}{B_v}. \qquad \text{(VII.1)}$$

Consider next the test rig in which the disturbance forces are carried wholly by the test bearing, and are superimposed on a steady load which is transmitted (wholly or partly) through reaction bearings. Assume the shaft rigid and the bearing rigidly supported, with excitation of disturbance forces provided by out-of-balance of the shaft.

The disturbance forces on the bearing represent the acceleration forces on the shaft; these arise jointly from the out-of-balance and the acceleration of the shaft axis. When the shaft axis moves in an elliptical orbit the disturbance forces no longer correspond to a vector of fixed magnitude rotating at shaft speed; the forces are not pure out-of-balance forces.

The acceleration of the centroid of the shaft is

$$u' - E'\omega^2 \cos \omega t \qquad \text{in the } u\text{-direction}$$

$$v' - E'\omega^2 \sin \omega t \qquad \text{in the } v\text{-direction}$$

where u', v' are actual displacements, and E' is the eccentricity giving out-of-balance.

Hence with F'_u, F'_v representing accelerating forces applied to shaft,

$$F'_u = M(u' - E'\omega^2 \cos \omega t)$$

$$F'_v = M(v' - E'\omega^2 \sin \omega t).$$

The steady load on the bearings is W gravitational units, or gW absolute units. On a rotor loaded only by its own weight $W = M$; but on the usual design of test rig, W is much greater than M.

The nondimensional forces transmitted from journal to bearing are defined by

$$P_u = -\frac{F_u'}{gW}, P_v = -\frac{F_v'}{gW}$$

and the nondimensional displacements by

$$u = \frac{u'}{r}, \quad v = \frac{v'}{r}.$$

Hence with $\Omega^2 = (Mr/gW)\omega^2$ and with $E = (E'/r)$, the equations of motion become

$$P_u + \Omega^2\left(\frac{\ddot{u}}{\omega^2}\right) = \Omega^2 E \cos \omega t$$

$$P_v + \Omega^2\left(\frac{v}{\omega^2}\right) = \Omega^2 E \sin \omega t.$$

Taking P_u, P_v as represented by the assumed dynamic characteristics,

$$\Omega^2\left(\frac{\ddot{u}}{\omega^2}\right) + B_u\left(\frac{\dot{u}}{\omega}\right) + K_u u = \Omega^2 E \cos \omega t$$

$$\Omega^2\left(\frac{\ddot{v}}{\omega^2}\right) + B_v\left(\frac{\dot{v}}{\omega}\right) + K_v v = \Omega^2 E \sin \omega t.$$

These relations can be satisfied only when

$$\frac{K_u - \Omega^2}{K_v - \Omega^2} = \frac{B_u}{B_v}. \qquad \text{(VII.2)}$$

References

[1] REYNOLDS, O., 'On the Theory of Lubrication', *Phil. Trans. Roy. Soc. A*, 1886, **177**, 157.

[2] *General Discussion on Lubrication*, I.Mech.E., 1937.

[3] HERSEY, M. D., *Theory and Research in Lubrication*, 1966 (John Wiley, New York).

[4] 'Lubrication and Wear: Fundamentals and Application to Design', *Proc. I.Mech.E.*, 1967, **182**, Part 3A.

[5] RAMSDEN, P., 'Review of Published Data and their Application to the Design of Large Bearings for Steam Turbines', 1967 I.Mech.E. Conf. on Lubrication and Wear, p. 75.

[6] TABOR, D., 'Solid Friction and Boundary Lubrication', 1967 I.Mech.E. Conf. on Lubrication and Wear, p. 262.

[7] Barwell, F. T. *Lubrication of Bearings*, 1956 (Butterworths, London).

[8] WILCOCK, D. F. and BOOSER, E. R., *Bearing Design and Application*, 1957 (McGraw-Hill, New York).

[9] PINKUS, O., and STERNLICHT, B., *Theory of Hydrodynamic Lubrication*, 1961 (McGraw-Hill, New York).

[10] CAMERON, A. and WOOD W. L., 'The Full Journal Bearing', *Proc. I.Mech.E.*, 1949, **161**, 59.

[11] OCVIRK, F. W., 'Short-bearing Approximation for Full Journal Bearings', *N.A.C.A. Tech. Note*, 2808, 1952.

[12] WILCOCK, D. F., 'Predicting Sleeve-bearing Performance', 1957 I.Mech.E. Conf. on Lubrication and Wear, p. 82.

[13] BATCHELOR, G. K., *An Introduction to Fluid Dynamics*, 1967 (Cambridge University Press).

[14] COLE, J. A., 'Experiments on the Flow in Rotating Annular Clearances', 1957 I.Mech.E. Conf. on Lubrication and Wear, p. 16.

[15] TAYLOR, G. I., 'Stability of a Viscous Liquid contained between two Rotating Cylinders', *Phil. Trans. Roy. Soc. A*, 1923, 289.

[16] TAYLOR, G. I., 'Fluid Friction between Rotating Cylinders', *Proc. Roy. Soc. A*, 1936, **157**, 546.

[17] DIPRIMA, R. C., 'Stability of Curved Flows', *Trans. A.S.M.E.*, 1963, **30**, Series E, 486.

[18] DIPRIMA, R. C., 'A Note on the Stability of Flow in Loaded Journal Bearings', *A.S.L.E. Trans.*, 1963, **6**, 249.

[19] DONNELLY, R. J. and SIMON, N. J., 'An Expirical Torque Relation for Supercritical Flows between Rotating Cylinders', *J. Fluid Mech.*, 1960, **7**, 401.

[20] WILCOCK, D. F., 'Turbulence in High Speed Journal Bearings', *Trans. A.S.M.E.*, 1950, **72**, 825.

[21] SMITH M. I. and FULLER, D. D., 'Journal Bearing Operation at Superlaminar Speeds', *Trans. A.S.M.E.*, 1956, **78**, 469.

[22] NEALE, M. J. and LOVE, P. P., 'High Speed Journal Bearings', 1957 I.Mech.E. Conf. on Lubrication and Wear, p. 123.

[23] HUGGINS, N. J., 'Tests on a 24-in.-diameter Journal Bearing: Transition from Laminar to Turbulent Flow', *Proc. I.Mech.E.*, 1966–7, **181**, Part 3B, Paper 1.

[24] DUFFIN, S. and JOHNSON, B. T., 'Some Experimental and Theoretical Studies of Bearings from Large Turbine Generator Sets', *Proc. I.Mech.E.*, 1966–7, **181**, Part 3B, Paper 4.

[25] NEWKIRK, B. L. and TAYLOR, H. D., 'Shaft Whipping due to Oil Action in Journal Bearings', *General Electric Review*, 1925, **28**, 559.

[26] LINN, F. C. and PROHL, M. A., 'The Effect of Flexibility of Support upon the Critical Speeds of High Speed Rotors', *Soc. Nav. Arch. & Mar. Engrs.*, Nov. 1951.

[27] MORRISON, D., 'Influence of Plain Journal Bearings on Whirling Action of an Elastic Rotor', *Proc. I.Mech.E.*, 1962, **176**, 542.

[28] SMITH, D. M., 'Dynamic Characteristics of Turbine Journal Bearings', 1963 I.Mech.E., Conv. on Lubrication and Wear, p. 72.

[29] SMITH, D. M., 'Journal-bearing Dynamic Characteristics–Effect of Inertia of Lubricant', 1965 I.Mech.E. Conv. on Lubrication and Wear, p. 177.

[30] DUFFIN, S. and GIBSON, W. H., 'Rig for Testing and Developing Large Turbine-generator Bearings', *The Engineer*, 1966, **222**, 785.

[31] HAGG, A. C. and SANKEY, G. O., 'Some Dynamic Properties of Oil-film Journal Bearings with Reference to Unbalance Vibration of Rotors', *J. Appl. Mech.*, 1956, **23**, 302.

[32] HAGG, A. C. and SANKEY, G. O., 'Elastic and Damping Properties of Oil-film Journal Bearings, for Application to the Unbalance Vibration Calculations', *J. Appl. Mech.*, 1958, **25**, 141.

[33] GLIENECKE, J., 'Experimental Investigation of the Stiffness and Damping Coefficients of Turbine Bearings', *Proc. I.Mech.E.*, 1966–7, **181**, Part 3B, Paper 13.

[34] SMITH, D. M., 'Vibrations in Turbomachinery', *Proc. I.Mech.E.*, 1966, **180**, Part 3I.

[35] HOTHER-LUSHINGTON, S., 'Failure of Turbine Bearings using Chromium-alloy Steel Journals', 1966 I.Mech.E. Conv. on Lubrication and Wear, Paper 11.

[36] DAWSON, P. H. and FIDLER, F., 'Wire Wool Type Bearing Failures: The Effect of Steel Composition, Structure and Hardness', 1967 I.Mech.E. Conf. on Lubrication and Wear, Paper 20.

[37] KARPE, S. A., 'Influence of Rotor Metal on Bearing Failures Generally Classified as the Machining Type', *Proc. I.Mech.E.*, 1967–8, **182**, Pt. 1, 203.

[38] *British Standard for Steam Turbine Oils*, B.S. 489 : 1968.

[39] FOWLE, T. I., 'Lubricants for Fluid Film and Hertzian Contact Conditions', 1967 I.Mech.E. Conf. on Lubrication and Wear, p. 568.

[40] MUYDERMAN, E. A., 'New Possibilities for the Solution of Bearing Problems by Means of the Spiral Groove Principle', 1966 I.Mech.E. Conf. on Lubrication and Wear, Paper 9.

[41] HIRS, G. G., 'Partly-grooved, Externally Pressurized Bearings', 1966 I.Mech.E. Conv. on Lubrication and Wear, Paper 21.

[42] TONDL, A., *Some Problems of Rotor Dynamics*, 1965 (Chapman & Hall, London).

[43] NEALE, M. J., 'Selection of Bearings', 1967 I.Mech.E. Conf. on Lubrication and Wear, p. 547.

[44] FORRESTER, P. G., 'Soft-Metal Bearings', 1967 I.Mech.E. Conf. on Lubrication and Wear, p. 321.

[45] LANCASTER, J. K., 'Composite Self-Lubricating Bearing Materials', *Proc. I.Mech.E.*, 1967–8, **182**, Part 1, 33.

[46] BEUERLAIN, P. and KARA, W. H., 'Selection of Lubricants', 1967 I.Mech.E. Conf. on Lubrication and Wear, p. 557.

[47] BLANCHARD, P. M., 'Lubricants for Hot Environments', 1967 I.Mech.E. Conf. on Lubrication and Wear, p. 472.

[48] WILSON, A. C. M., 'Fire-Resistant Fluids for General Hydraulic and Steam Turbine Systems', *Proc. I.Mech.E.*, 1967–8, **182**, Part 1, 109.

[49] HOTHER-LUSHINGTON, S. and SELLORS, P., 'Water Lubricated Bearings: Initial Studies and Future Prospects in the Power Generation Industry', 1963 I.Mech.E. Conv. on Lubrication and Wear, p. 139.

[50] HOTHER-LUSHINGTON, S., GARSIDE, D. W., and SELLORS, P., 'Water Lubricated Bearings – Further Experiments', 1964 I.Mech.E. Conf. on Lubrication and Wear, p. 89.

[51] GARSIDE, D. W. and HOTHER-LUSHINGTON, S., 'The Influence of Clearance and Journal Surface Finish on the Lead Capacity of Water Lubricated Plain Bearings', 1966 I.Mech.E. Conv. on Lubrication and Wear, Paper 14.

[52] SAMUELSON, F., 'Lubrication of Journal Bearings with a Water-base Lubricant', General Discussion on Lubrication, *Proc. I.Mech.E.*, 1937, **1**, 269.

[53] GROSS, W. A., 'Gas Bearings – Journal and Thrust', 1967 I.Mech.E. Conf. on Lubrication and Wear, p. 116.

[54] SPIERS, H. M., *Technical Data on Fuel*, 1955 (British Nat. Comm., World Power Conf., London).

[55] CAMERON, A., *The Principles of Lubrication*, 1966 (Longmans, Green, London).

[56] *A Guide to the Viscosity of Liquid Petroleum Products*, 1967, Engg. Sciences Data Item No. 67015 (I.Mech.E., London).

[57] *Information on Factors affecting the Selection of Lubricants*, PD6428, 1969 (British Standards Institution).

Index

In this index, the word 'bearing' usually implies 'journal bearing'.